EcoProduction

Environmental Issues in Logistics and Manufacturing

Series editor

Paulina Golinska-Dawson, Poznań, Poland

The EcoProduction Series is a forum for presenting emerging environmental issues in Logistics and Manufacturing. Its main objective is a multidisciplinary approach to link the scientific activities in various manufacturing and logistics fields with the sustainability research. It encompasses topical monographs and selected conference proceedings, authored or edited by leading experts as well as by promising young scientists. The Series aims to provide the impulse for new ideas by reporting on the state-of-the-art and motivating for the future development of sustainable manufacturing systems, environmentally conscious operations management and reverse or closed loop logistics.

It aims to bring together academic, industry and government personnel from various countries to present and discuss the challenges for implementation of sustainable policy in the field of production and logistics.

More information about this series at http://www.springer.com/series/10152

Menka Khoobchandani · Arpita Saxena
Editors

Biotechnology Products in Everyday Life

Springer

Editors
Menka Khoobchandani
Department of Radiology
University of Missouri
Columbia, MO, USA

Arpita Saxena
AAPT Foundation
Aurangabad, India

ISSN 2193-4614 ISSN 2193-4622 (electronic)
EcoProduction
Environmental Issues in Logistics and Manufacturing
ISBN 978-3-319-92398-7 ISBN 978-3-319-92399-4 (eBook)
https://doi.org/10.1007/978-3-319-92399-4

Library of Congress Control Number: 2018955453

This Springer imprint is published by the registered company Springer Nature Switzerland AG
The registered company address is: Gewerbestrasse 11, 6330 Cham, Switzerland

We dedicate this effort to our parents, gurus, & to budding entrepreneurs who aspire to better the world with new ideas....

Foreword from Academia

Biotechnology is dynamically updating and upgrading its horizons. Students—teachers and researcher communities—need to keep themselves abreast with the constantly changing dynamics of this subject. Dr. Arpita Saxena and Dr. Menka Khoobchandani are committed to this change and have put their best foot forward through this book—*Biotechnology Products in Everyday Life*. The book covers a spectrum of 16 reviews and research articles categorized in seven parts. These parts are artistically named after colours of rainbow VIBGYOR that symbolize variety and beauty that the book offers—ranging from nanoparticle history and present medicinal applications in ailments like cancer. The flow drives us towards intelligent devices that act as interface between humans and machines, speeding up recovery by complementing medical practices or acting as alternatives altogether. The book revisits a long-discussed topic—bioplastics—and talks about green energy, removal of arsenic from groundwater up to 97% in order to make it potable as per WHO guidelines, using solar energy. Moving ahead, we come across a point where this has the audacity to reconsider the status of genetically modified crops as well as proposes orphan crops guaranteeing food and nutrition security.

Having discussed about the crops which ensure 'food for all', the book takes us ahead to improvement of crop production of popular crops like soya bean by novel metabolites. Interestingly, the book also touches the applications of phyto-stem cells in cosmetic industry, which catches the limelight often. These products having stem cells capture the customers' attention for being topically safe, elevating the skin quality and their anti-ageing properties. The next segment demonstrates an orchestrated effort of medical faculties to cure cerebral cavernous malformations, by targeted pharmacotherapies. Another author in the same section writes about current trends on herbal neuro-protection and how plant-derived pure compounds/extracts are preferred alternatives to treat nervous system disorders because of their non-toxic and long-lasting nature. Cancer is the most dreaded disease today, and targeted action against these cells with lost/mutated set of normal genes is most sought after. An article on dual targeting of cancer cells by radiolabeled metallic and polymeric nanoparticles is intriguing and reliving at the same time. The flow of ideas that started with nanoparticles finds another abode, where the cancer cells are countered by devices having intelligent targeted systems using nanoparticles.

The courageous ideas collaged by the authors from various locations across globe is a fresh change that transforms the mindsets of young readers towards broader spectrum of application of their subject. Spearheading one of the leading universities of India gives me the clarity to say that all research must lead to products. Products those are useful, competitive, commercially viable and scientifically bright. I have taken many initiatives to encourage entrepreneurship amongst my students including setting up an incubation centre for the excellence in innovation and entrepreneurship. We believe that academic non-compartmentalization should be encouraged and multidisciplinary projects leading to effective change should be brought about in the education system at all levels.

With the wide scope the book offers, I see many collaborative ideas and innovative products on the way. I congratulate Dr. Arpita Saxena and Dr. Menka Khoobchandani for successful editing of this book.

Chicago, USA

Prof. B. A. Chopade, Ph.D.
(Microbiology and Molecular
Genetics) Nottingham University
Fogarty Fellow (Genetic Engineering)
Illinois University

Aurangabad, Maharashtra, India

Present Position-Vice Chancellor
Dr. Babasaheb Ambedkar
Marathwada University

Foreword from Industry

Biotechnology has the key to revolutionize the future to a vast extent. To understand this in a better way, the editors of this book, Dr. Menka Khoobchandani and Dr. Arpita Saxena, have laid down simple and understanding modules to explain the upcoming technological advances through their unique concept of VIBGYOR of biotechnology.

This book opens mind to understand the current developments going in each field and explains the reason how biotechnology is the key in this major breakthrough.

The authors have openly shared their developments in biotechnology and how the future would be with the possible discoveries. The authors are reputed and have immense scientific knowledge by which they have made the subject be easy to understand.

This book is going to have a lasting impression for all people who love biotechnology and are keen to expand their knowledge in research give them clarity moving forward.

This is truly a Stunner Recommendation, and the word VIBGYOR each letter is well coined and more importantly capturing the flow of the topics for better understanding of the book.

The knowledge of medicine has changed dramatically over the years through pioneering advances in biotechnology research and innovation.

Chennai, India

Mr. S. Abhaya Kumar Jain
Founder of Shasun Pharmaceuticals
and LifeCell International (Leading
Pharmaceutical and Stem Cell
Banking Company, India)

Mr. Deepak Abhaya Srisrimal
Dhanvantari Nano Ayushadi Pvt Ltd.

Preface

Stories are loved by young and old alike. Stories of human evolution, of correlating 'Mendelian factors' with genes, of coining the term 'mutation' by Hugo De Vries, of developments related to the fields of medicine, agriculture and technology and contemporary stories of entrepreneurs who made it big in their own and others' lives because of their ideas. Therefore to start with, we would first narrate the story of conception of this book.

While pursuing our Ph.D.s, we met in 2010 when Menka came for a few days' training at Indian Institute of Integrative Medicine, Jammu, India, where Arpita was doing her Ph.D. Our intellectual matches caught each other like ligases and we started discussing personal and professional lives often. Even after completion of our Ph.D.s, we were in touch. While Menka moved to the University of Missouri, USA, for her postdoctoral job, Arpita settled in Aurangabad, India, to have a family and started working with a start-up. We often discussed with each other the life of a researcher and differences/similarities in USA and India. We felt the challenges are more or less the same everywhere and that a fair exposure, leaving comfort zones and moving to new places not only improves our professional outlook but also refines personal mindsets.

April of 2017, Arpita was thinking about how biotechnology as a subject was projected in the year 2000 in India, and what its present scope in research was. She shared the thought of writing/editing a book on biotechnology for youngsters, to help them decide what to exactly expect from this field of science. Menka could not agree more and meticulously she got on board authors who would sync in with the idea and aim to bring innovative scientific ideas from complicated journals into a simple book form. The journey became more and more interesting as we saw enthusiasm and energy from our all authors. The team spirit reflects the 'We believe magic should be felt' theory.

This book is written with a not-too-technical narrative aimed at undergraduates and graduates broadly covering the application of biotechnology products in everyday life without sacrificing its deeper understanding. As a result, it should have broad interest for use in college course on biotechnology, bioscience, biology, material sciences and bioengineering. While this book does not necessarily challenge current trends in the field of biotechnology, it contributes to the ongoing research over the nature and causes of this sector.

Innovations in biotechnology have kept the promise of being ‘biological-technical- and more importantly-useful’ to every bit of the word essentially for the scope of enactment it gives to the researchers and the ways it brings relief to humans. We have also attempted here to build bridges between biotechnology and intrarelated subjects like drug development, environmental sciences, botany and agricultural sciences, and various other interrelated disciplines like nanotechnology, material sciences and robotics, which is relevant in today’s world of interdependence. There has been a shift in work culture throughout the globe, and people are turning towards enterprising their ideas instead of engaging in academic research. Where on the one hand, both are equally important for the complete understanding of the subject, somewhere this generation takes pride in being the job creators rather than being job seekers.

This book celebrates various interdisciplinary approaches to address man’s biotechnological quests. This is divided into seven categories, each containing 1–3 chapters. The chapters are in the form of colours, which are visible to human eye while triggering the readers to explore those which are not!! The reviews are simplistic and colourful. Deep and most advanced research is woven into VIBGYOR.

Very small world: The first category focuses on the role of nanotechnology and nanomaterials from ancient time to present times; and nanosizing drugs as latest drug delivery strategy. In this category, authors write about safe and modern application of green biotechnology that could be useful for cancer cure and early detection.

Intelligent devices: The second category explores the devices incorporated with artificial intelligence. It also reviews various products born as result of marriage of computer sciences to bioengineering while also showcasing point of care devices and smart devices such as the pacemaker.

Behind the Scenes Genes: Here the authors share the current trends on CAR-modified T-cell therapy, which is a highly promising treatment for cancer. Here T cells are engineered to express the antigen-specific CAR to be injected into the patients to cure cancer.

Greener Environment: As we move towards industrialization, pollution is faced as the biggest roadblock, and if the groundwater is polluted, the situation just gets worst. In this category, the authors propose methods to purify arsenic contamination from groundwater and also talk about bioplastics.

Yielding more (agricultural and industrial): This category deliberates about the status of genetically modified crop (GMO), which is a classic feature of agricultural biotechnology. At the same time, it converses about bringing orphan crops to the market; in order to fill the gaps and snags for most of the identified species as no concrete scientific data is globally available.

Ordinary to extraordinary: Metabolomics is one of the most emerging technologies being used for crop improvement. Here analytical chemists step in with integrated approaches incorporating genomics, transcriptomics, proteomics, ionomics and metabolomics for soybean improvement. Another chapter in this category elaborates on plant stem cells and their use in cosmetology.

Readiness to cure: The applications in medical and pharmaceutical industry are the most indispensable contribution of biotechnology. This category takes us to three diverse topics ranging from cancer to neural disorders to cerebral cavernous malformations and how biotechnology equips us to deal with these ailments better.

We, however, could not bring every topic related to the subject, but we picked up the most trending and impacting ideas. "Does the highest end of my expectation get fulfilled here, or this book still triggers me to think more/different in related spaces?" is the question this book leaves its readers with. Regardless of the last thought of the reader, this book is a very useful experience for the readers to channelize their minds with a non-academic temper, atypical for the youngsters and make them approach the subject in a more enterprising and non-compartmentalized manner which is a prerequisite for a successful research career.

Wishing you a happy read!

Aurangabad, India Arpita Saxena
Columbia, USA Menka Khoobchandani

Contents

About the Editors

Dr. Menka Khoobchandani is a research scientist within the Department of Radiology and the Institute of Green Nanotechnology at the University of Missouri Medical School in Columbia, Missouri, USA, having 10 years of research and academic experience. She received her Ph.D. in Chemistry (natural product and cancer biology research areas) from Dayalbagh Educational Institute, India. Her research focuses towards the interdisciplinary areas of sciences—nanotechnology, nanomedicine, cancer biology, drug discovery and material sciences. Dr. Khoobchandani is an active collaborator with Dhanvantari Nano Ayushadi Private Limited (DNA), Chennai- India developing tumor specific nanoparticles and screening them using in vitro and in vivo assays to establish their efficacy in treating various tumors. Dr. Khoobchandani's vision is to create an innovative research environment, to produce safe therapeutics methods that can lead to the development of alternative technologies towards the unmet clinical needs. She has established global collaborative partnerships with professors and clinicians across universities and hospitals, resulting in joint projects, patents and publications.

Arpita Saxena is a social entrepreneur, striving to better the social education system in India with her concept of AAPT. Arpita finished her Ph.D. in biotechnology in 2012 off campus from Guru Nanak Dev University, Amritsar, India, while working on Anticancer therapeutics at the department of Cancer Pharmacology, Indian Institute of Integrative Medicine (CSIR), Jammu, India.

Later she was working with a start-up at Aurangabad, funded by BIRAC, India, for 2 years. Arpita is associated with BYST (a not-for-profit organization to encourage entrepreneurship amongst Indian youth) and recognized by City and Guilds for mentoring skills. Teaching and writing continue to be her motivation.

Part I
Very Small World (Nanodeliveries)

Nanotechnology in Ancient Era

Ketaki Deshmukh

Abstract Nanotechnology is defined as design and production of structures, devices, and systems by controlled manipulation of sizes and shapes at atomic, molecular, and macromolecular scale where properties differ significantly from those of bulk materials. Although the concept of "*nanoscience/nanotechnology*" was introduced by the Nobel laureate Richard Feynman in 1959, long before the beginning of "nanoera," ancient people used various nanoparticles and processes related to it. Some classic examples are the Lycurgus cup: A cup made up of glass containing gold–silver-alloyed nanoparticles which change color from greenish-yellow to red when light is shone on it; Maya blue: a corrosion-resistant blue pigment which consists of indigo molecules incorporated into needle-shaped palygorskite crystallites; Damascus steel sword: an unbreakable and exceptionally sharp sword built using steel blades containing oriented nanowires- and nanotube-like structure. Nevertheless, the breakthrough of nanotechnology has been permitted over the last three decades due to invention of many techniques/instrument which allowed the manipulation and observation of the nanoworld. Nowadays, nanomaterials/nanoparticles are being used for many applications in daily life, such as in the fields of electronics, catalysis, optics, biology, and medicine. This chapter presents an overview of nanotechnology, from ancient eras.

Keywords Nanotechnology · Ancient era · Nanoparticles and nanomaterials

1 Introduction

The manipulation of material at the atomic and molecular scale to create new functions and properties sounds like it should be a profoundly modern concept. However, from the ancient era, ancestors were controlling the matter at the atomic scales. By modern-day standards, they were working in a branch of engineering called nanotechnology [1].

K. Deshmukh (✉)
Prollenium Medical Technologies Inc., Toronto, Canada
e-mail: ketaki03@gmail.com

M. Khoobchandani and A. Saxena (eds.), *Biotechnology Products in Everyday Life*, EcoProduction, https://doi.org/10.1007/978-3-319-92399-4_1

The concept of "*nanoscience/nanotechnology*" was introduced by the Nobel laureate Richard Feynman in 1959, in his visionary talk entitled "*There is plenty of room at the bottom*" at the American physical society meeting. Though he did not explicitly mention the term "*nanotechnology*", he suggested the eventual possibility of manipulating atoms and molecules precisely in a desired fashion to create molecular machines. According to him, the advent of this technology could facilitate the writing of the entire 24 volumes of the Encyclopedia Britannica on the head of a pin [2]. In 1974, the term "*nanotechnology*" was used for the first time by Professor Norio Taniguchi to describe the processes of creating semiconductor with nanometer dimension. The term was used again in 1981 when Eric Drexler, inspired by Richard Feynman's talk, developed and popularized the concept of nanotechnology [3]. He published the famous book "Engines of creation: the coming era of nanotechnology" in which he used the word nanotechnology to describe engineering on the billionth of a meter scale. Invention of scanning tunneling microscope and the atomic force microscope has offered opportunities for scientists to the "nanoworld" by providing them the tools not only to image surfaces with atomic resolution, but also to move individual atoms as previously predicted by Richard Feynman [4].

Nanotechnology seems to be associated with modern science, it must be noted that long before the beginning of "nanoera" use of various nanoparticles and processes related to it have been performed. Indeed, researchers have analyzed a lot of antiquities with unique characteristics such as unusual colors, strength, and anticorrosion properties that have been attributed to the presence of nanoparticles [1, 5]. Evidences of use and applications of various nanoparticles such as gold, silver, copper, and carbon can be traced back to the time period of BCE. Some historic examples are the Lycurgus cup invention of Romans; Maya blue: a corrosion-resistant blue pigment, Damascus steel sword: an unbreakable and exceptionally sharp sword. Nanotechnology was not only limited to artifacts, it was also used in ancient medicine in the form of Bhasmas. The Ayurvedic medicine "Bhasmas" is the most ancient applications of nanomedicine. This chapter presents an overview about applications of nanotechnologies in ancient era [1, 5].

2 Classic Examples of Ancient Nanotechnologies

2.1 Lycurgus Cup

The most famous example of the use of metallic nanoparticles in ancient era, a stunning decorative Roman glasswork from about AD400, the Lycurgus cup which shows a mythological frieze depicting the legend of King Lycurgus (Fig. 1). The mythological scenes on the cup illustrate the death of Lycurgus, King of the Edoni in Thrace at the hands of Dionysus and his followers. Lycurgus attacked Dionysus and Ambrosia, one of his maenads. Ambrosia called out to Mother Earth, who transformed her into a vine. She then coiled herself on the king and captivated him for his evil behavior.

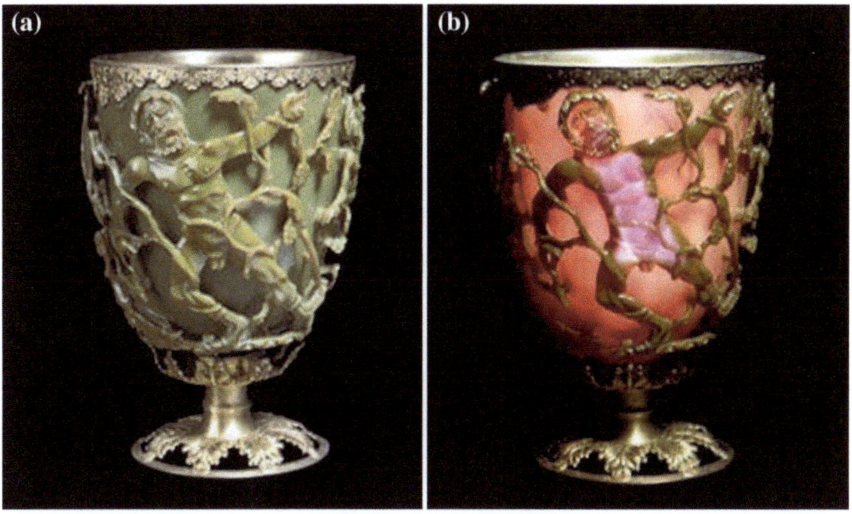

Fig. 1 Lycurgus cup in reflected (**a**) and in transmitted (**b**) light [6]

The cup portraits this moment of Lycurgus entangled in vines. Lycurgus cup is not only famous for its decorative artwork but also for unusual optical effects. The most remarkable aspect of this cup relies on its dichroism: it appears jade green when lit from the front but blood-red when lit from behind (Fig. 1a, b) [6].

In 1958, the British Museum acquired the cup from Lord Rothschild (with the aid of a contribution from the National Art Collection Fund) [6]. In early modern age, this cup first came to scholarly attention and preliminary studies including qualitative spectrographic analysis were carried out at the British Museum and sample was sent to General Electric Company Ltd. (GEC) at Wembley for more detailed microanalysis to discover the reason behind unusual optical effects [7]. B. S. Cooper at GEC reported that the presence of trace quantities of gold, silver, and other elements in the glass and further suggested that the unique optical characteristics of the glass might be connected with the presence of colloidal gold in the glass [6, 8]. To further understand the remarkable color effect, sample was sent to Dr Robert Brill of the Corning Museum in 1962. Brill in collaboration with GEC carried out detailed microanalysis on the Lycurgus cup and confirmed that the dichroism was linked to the presence of minute amounts of both gold (about 40 ppm) and silver (about 300 ppm) in the glass [9, 10]. However, simply adding traces of gold and silver during formulation of glass would not show this unique optical property. Therefore, Brill suggested that during the glass manufacture, these metallic salts have been probably added in the silicate and then partially reduced with suitable reducing agents during further heat treatment of the glass. Thus, the different colors of the glass are attributed to the presence of this metallic colloidal dispersion, in particular to a combined effect of the absorption of gold which leads to the red transmission, and the scattering of silver which leads to the green reflectance [5, 6, 9, 10]. This dichroism phenomenon

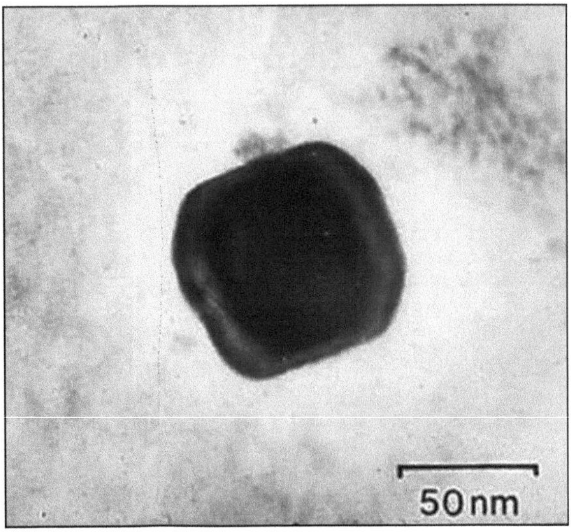

Fig. 2 Transmission electron microscopy (TEM) image of a silver–gold alloy particle within the glass of the Lycurgus cup. The Trustees of the British Museum [6]

has been referred as the Lycurgus effect [5, 6, 9, 10]. The process used for making Lycurgus cup was not well understood or controlled by the makers and was probably accidental discovery. It could be attributed to "contamination" with minutely ground gold and silver dust. The glassmakers may be unaware of the fact that gold and silver were involved, as the quantities were minute. Due to lack of technology, Brill was not able to prove the presence of metallic particles and the relative contributions of silver and gold to the optical effect. Therefore, in the late 1980s, a further small fragment of the cup was examined by Barber and Freestone [11]. Thus, to further analyze, these metal particles were observed under transmission electron microscopy and to their surprise it was observed that these particles were nanoparticles with diameter of 50–100 nm (Fig. 2) [9–11]. In addition to this, X-ray diffraction analysis has shown that these nanoparticles are silver–gold alloy, with a ratio of silver to gold of about 7:3, containing in addition about 10% copper [9–11]. The Lycurgus cup is the outstanding example of Roman technology in every aspect; its excellent cutwork and unique red-green dichroism render it a distinctive record.

2.2 *Stained Glass Windows in European Cathedrals During Medieval Periods*

Stained glass window is another classic example of use of gold and silver nanoparticles during ancient era. In the Medieval age, many churches in Europe were decorated with bright and vibrant colored stained glass windows. Colors were historically related to emotions; rich red and glorious yellow were traditionally used in stained glass. Many people were illiterate, and also books were rare during this time; there-

Fig. 3 Stained glass window from the Sainte-Chapelle in Paris Centre des Monuments Nationaux [5]

fore, stained glass windows played educational role by narrating stories through pictures [5] (Fig. 3).

The glass was manufactured by melting pure sand (SiO_2), which required very high temperature of about 2500 °F. Therefore, to reduce temperature, artisans mixed sand with soda ash, lime, potash, and lead oxide, which reduced temperature to 1500 °F. After melting, coloring agents, or colorants, were added [12, 13]. Various colors attributed to different chemical compounds that were added to the molten glass during processing. In some cases, the colorants were part of the basic glass making process (i.e., impurities found in the sand were used to make the glass, or from smoke generated in the firing process). Artisans reported that various metallic compounds (typically metal oxides, sulfides, and chlorides) were responsible for different colors [5, 14]. Recent analysis by scientist showed that adding gold chloride and silver nitrate gave red and yellow color, respectively, were originated from colloidal gold and silver nanoparticles formed during the glass manufacturing process [12, 13]. Particle-induced X-ray emission analyses have revealed the presence of gold with a concentration between 10 and 35 ppm [5, 12, 13]. The artisans were unaware of the fact that the color they observed after adding gold chloride and silver nitrate were from nanoparticles of gold and silver created during the glass process [14].

Use of Nanotechnology in Ancient Cosmetic Cosmetics have been used for thousands of years and were mostly made using the mixture of various naturally available oils, plant extract, and water. Use of nanoparticles in sunblock, toothpaste, and

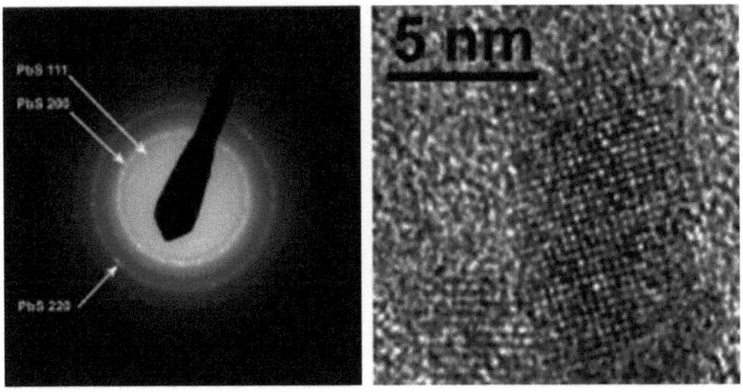

Fig. 4 Observation and identification by electron microscopy (HRTEM) of PbS crystallites inside the cortex. **a** Electron diffraction pattern: the d spacing is consistent with the unit cell of PbS. **b** High-resolution micrograph of a representative PbS nanocrystal [16]

cream sounds like a new technology; however, ancient Greeks and Romans were using nanotechnology in their cosmetics. Ancient Greeks and Romans were not only using cosmetics formulated using naturally available plant extracts but they also had knowledge of chemical formulation of cosmetics using materials such as lead white (lead carbonates) as foundation cream and galena for eye makeup [15, 16]. Since the Greco-Roman period, dying hair was a very common beauty treatment and they were using organic hair dyes obtained from plants such as henna and also other unusual recipes based on lead compounds [17]. A Greco-Roman hair-dye recipe based on lead compound involves applying a paste of lead oxide and calcium hydroxide, or lime, to greying hair. The lead-lime blend reacts with sulfur present in hair's keratin proteins and forms crystals of lead sulfide, also called galena which gives rise to the black color [16]. This process of dying hair was developed 2000 years ago and also been used in modern era.

Scientists were interested to know the size of those penetrating lead sulfide crystals; therefore, they soaked blond hairs in a solution of the Greco-Roman chemicals for up to three days. Upon sufficiently darkening of the strands, the research team examined it under an electron microscope and they observed that the strands were covered with lead-sulfide crystals of averaging 4.8 nm in size (Fig. 4). It is remarkable that Romans-Greeks were using nanotechnology in the dyeing process and were aware of the basic chemistry methods to develop low-cost methods [15, 16].

Maya Blue Mayan archeological sites always surprised scientists due to the blue color often used in pottery, murals, and ceremonial artifacts. Unlike other blue color identified in ancient paintings, this Mayan blue color does not contain copper, lapis lazuli, or lazurite [18]. In the late twentieth century, Maya blue was used in Mesoamerica and colonial Mexico. In addition to its beautiful color, it is resistant to biodegradation, acids, alkalis, high humidity, and temperatures [19–21]. The Bonampak archeological (Fig. 5) site is a classic example of Mayan painting in which

Fig. 5 Use of Maya blue pigment in the famous wall paintings at Cacaxtla, Mexico

blue color has not faded even after centuries in the extreme conditions of the rain [19, 20]. The chemical structure and the origin of the Maya blue color have been debated extensively among researchers. It has been reported that it mainly contains clay (mainly palygorskite mixed with smaller amounts of sepiolite and montmorillonite) mixed with indigo dye which was obtained from the sprigs of a plant called xiuquilit (*Indigophera* sp.) [22]. Reason behind anticorrosion property of Maya blue was unclear to the researchers for many centuries. Analysis of authentic samples of Maya blue using high-resolution transmission electron microscopy, electron energy loss spectroscopy, and x-ray microanalysis studies revealed that it contains needles of palygorskite and nanoparticles supported in an amorphous SiO_2 substrate. It mainly contains iron nanoparticles, with a minor quantity of chromium, titanium, and manganese. The SiO_2 substrate contains sodium and magnesium impurities. It has been observed that palygorskite and indigo molecule forms a superlattice and iron, magnesium, and copper metal nanoparticles encapsulated in silicate and oxide nanoparticles on the surface [19, 20, 23]. The anticorrosion property and beautiful tone of the color is obtained only when both the nanoparticles and the superlattice are present [19].

Scientist are able to synthesize Maya blue pigment with all the properties of historical Maya Blue in the laboratory; however, the exact technique that the Mayan used to synthesize such a sophisticated paint remains unknown [22]. Blue pigment is a complex of organic/inorganic material which contains clay with nanopores loaded with indigo to create an environmentally stable pigment. Furthermore, recently many researchers have been able to replace the organic compound, indigo with innumerable alternate organic dyes yielding novel dyes with various colors [20].

Damascus Steel Sword Damascus blades were first manufactured in the near East by using ingots of wootz steel which was supplied by India and Sri Lanka [24]. Assad Ullah, an Arabian, introduced the wootz steel, and sword is named after the capital city of Syria, Damascus. These 'Damascus blades' were strong, but flexible enough to bend from hilt to tip (Fig. 6). It was exceptional weapons that gave

Fig. 6 Eighteenth-century Persian-forged Damascus steel sword

the Muslims a benefit, and their blacksmiths succeeded in guarding the secret of its manufacturing. The secret eventually died out in the eighteenth century, and no European smith was able to fully reproduce this method [25].

The wootz steel was used to prepare Damascus blades which consist of iron with carbon that helps in hardening the metal. However, the problem with this steel was that high carbon contents that made the material harder and brittle. Therefore, wootz steel with its especially high carbon content of about 1.5% was useless for sword steel and the blade would shatter upon impact with a shield or another sword. On the other hand, the existing sabers showed a seemingly impossible combination of hardness and malleability [27].

In the twenty-first century, finally, researchers from Germany revealed the extraordinary secret of this sword which was based on the use of nanotechnology [26]. They solved this mystery by chemical and physical analysis of a Damascus sword manufactured by the famous blacksmith Assad Ullah. They dissolved part of the weapon in hydrochloric acid and studied it under an electron microscope. Interestingly, they observed that the sword contained carbon nanotubes; each one is half a nanometer larger than other [28]. These carbon nanotubes were cylindrical and hexagonally arranged. It was among the strongest materials with high elasticity and tensile strength [25]. In the analysis, it was found that the carbon nanotubes were protecting hard and brittle nanowires of cementite (Fe_3C); this structure helped blacksmiths to create strongest weapon [26, 28]. The malleability of the carbon nanotubes is due to the brittle nature of the cementite formed by the high-carbon containing wootz steel. It is not very clear to researcher about exact procedure used by ancient blacksmith for creation of these nanostructures. Researchers believed that small traces of metals such as vanadium, chromium, manganese, cobalt, and nickel were present in the wootz [24, 25]. These impurities were segregated during the hot and cold phases of making of word, and it would have acted as catalysts for the formation of the carbon nanotubes, which in turn would have promoted the formation of the cementite nanowires [26, 27]. One of the reasons that nobody was able to reproduce the sword could be the absence of particular combination of metal impurities present in the wootz steel [26–28]. However, it is interesting to know that nanotechnology was being used at least 400 years before it became the scientific buzzword of the twenty-first century.

Fig. 7 Bhasmas—Ayurvedic nanomedicine of ancient times

2.3 *Ayurvedic Bhasma: Nanomedicine*

Ancient scientist were not only using nanotechnology to create artifacts but was also popular in medicine. Ayurvedic Bhasma was an excellent innovation of ancient nanomedicine. They are unique Ayurvedic metallic/minerals preparation, mixed with herbal extract and were widely used by Indians for the treatment of various chronic disorders, since the seventh century (A.D.) [29]. It is an ash achieved through incineration; the starter material undergoes an elaborate process of purification followed by the reaction phase, which involves incorporation of some other minerals and/or herbal extract (Fig. 7). Interesting aspect of this procedure is that it is performed mechanically and not chemically [30]. Various herbal extract, metals, and non-metals preparations are used as medicine in Ayurveda and are prepared using Putapaka and Kupipakwa methods [31, 32].

Putapaka Method In this method, metals or minerals are heated at high temperature and then quenched in suitable media herbal juices or decoction for specified times and repeated this step to remove toxic effect of metals; this is called "Shodhana" (purification), and it is transformed into Bhasmas, the biologically active nanoparticles [30–32].

Kupipakwa Method Bhasma is prepared by introducing metals like gold, silver, and copper into Shodhana, Kajjali preparation, and Kupipaka. After purification, i.e., Shodhana process, metals or minerals are mixed with mercury and purified sulfur. Subsequently, the mixture is triturated till it formed black, lusterless, fine, and smooth paste known as Kajjali. Kajjali is dried, filled in a glass bottle called Kachkupi, covered by seven layers of cloth smeared with mud and then introduced into sand bath (Valukayantra) for homogeneous and indirect heating. Then the bottle is allowed to cool down, sublimed product is collected from the bottle and then ground to powder form [23, 30–33]. X-ray diffraction, TEM, and particle size analysis revealed that these Bhasmas are in nanometer dimension [30].

These processes reduce not only reduces the particle size to nanometer but also convert Bhasmas into biocompatible, bio-assimilable, absorbable, and suitable form for the human body. Various Bhasmas such as Swarna (gold), Rajat (silver), Makshika (pyrites), Abhrak (mica), Tamra (copper), and Louha (iron) are being used to treat many chronic diseases. The therapeutic effect of Bhasma may be attributed to large surface area of materials and small particle size by which they can easily transported into cell nucleus and to specific target sites [34, 35].

In the ancient time, this was a serendipitous discovery that heating of metals like gold, silver, copper, mercury, lead, iron, arsenic several times with herbal extract or other medicinal agents drastically alters the effectiveness of Ayurvedic medicines without producing any toxic or harmful side effects [32]. The scientist of ancient times had limited means to study that continuous heating and cooling of metals/minerals (in some case which was more than 100 times) could alter the chemical and physical properties of the parent metal [33]. They were performing this process to reduce the harmful effect of the metals and minerals; however, the outcome was incredible. The drugs resulted from this procedure were effective even at low dosage and also had longer shelf life. Therefore, nanometals and nanominerals were used routinely in Ayurvedic therapeutics and became strength of it [30, 33].

3 Conclusion

Nanotechnology is a fascinating branch of science, and it offers uncountable opportunities for surprising discoveries in various fields such as electronics, physics chemistry, and biology. It appears to be a new technology; however, it had already been used since ancient era. Use of various nanoparticles such as gold, silver, copper, and carbon has been demonstrated in the ancient era. Lycurgus cup, Maya blue, and Damascus sword are excellent creation of ancient times. Ancient scientist were not only using nanotechnology to create artifacts but were also popular in medicine. Ayurvedic Bhasmas are classical examples of nanomedicine. These unique innovations inspired many scientists to design new technologies.

Conflicts of Interest The author declares no conflict of interest.

References

1. Heiligtag, F.J., Niederberger, M.: The fascinating world of nanoparticle research. Mater. Today **16**(7–8), 262–271 (2013)
2. Feynman, R.P.: There's plenty of room at the bottom. Eng. Sci. **23**(5), 22–36 (1960)
3. Whitesides, G.M.: Nanoscience, nanotechnology, and chemistry. Small **1**(2), 172–179 (2005)
4. Bhushan, B.: Introduction to nanotechnology. Springer Handbook of Nanotechnology, pp. 1–19. Springer, Berlin, Heidelberg (2017)

5. Schaming, D., Remita, H.: Nanotechnology: from the ancient time to nowadays. Found. Chem. **17**(3), 187–205 (2015)
6. Freestone, I., Meeks, N., Sax, M., Higgitt, C.: The Lycurgus Cup—a Roman nanotechnology. Gold Bull. **40**(4), 270–277 (2007)
7. Whitehouse, D.: Roman glass in the Corning Museum of Glass. Hudson Hills **3**, 1–43 (1997)
8. Harden, D.B., Toynbee, J.M.: VII. The Rothschild Lycurgus Cup. Archaeologia **97**, 179–212 (1959)
9. Chirnside, R.C.: The Rothschild Lycurgus Cup: an analytical investigation. Proc. 7th Int. Cong. Glass **2**(222), 1–6 (1965)
10. Brill, R.H.: The chemistry of the Lycurgus Cup. Proc. 7th Int. Cong Glass. **2**(223), 1–13 (1965)
11. Barber, D.J., Freestone, I.C.: An investigation of the origin of the colour of the Lycurgus Cup by analytical transmission electron microscopy. Archaeometry **32**(1), 33–45 (1990)
12. Delgado, J., Vilarigues, M., Ruivo, A., Corregidor, V., Da Silva, R.C., Alves, L.C.: Characterisation of medieval yellow silver stained glass from Convento de Cristo in Tomar, Portugal. Nucl. Instrum. Methods Phys. Res. Sect. B **269**(20), 2383–2388 (2011)
13. Verità, M., Santopadre, P.: Analysis of gold-colored ruby glass tesserae in Roman church mosaics of the fourth to 12th centuries. J. Glass Stud. **1**, 11–24 (2010)
14. Sciau, P.: Nanoparticles in ancient materials: the metallic lustre decorations of medieval ceramics. In the delivery of nanoparticles. InTech **25**, 525–540 (2012)
15. Walter, P., Martinetto, P., Tsoucaris, G., Brniaux, R., Lefebvre, M.A., Richard, G., Talabot, J., Dooryhee, E.: Making make-up in ancient Egypt. Nature **397**(6719), 483–484 (1999)
16. Walter, P., Welcomme, E., Hallégot, P., Zaluzec, N.J., Deeb, C., Castaing, J., Veyssière, P., Bréniaux, R., Lévêque, J.L., Tsoucaris, G.: Early use of PbS nanotechnology for an ancient hair dyeing formula. Nano Lett. **6**(10), 2215–2219 (2006)
17. da França, S.A., Dario, M.F., Esteves, V.B., Baby, A.R., Velasco, M.V.: Types of hair dye and their mechanisms of action. Cosmetics **2**(2), 110–126 (2015)
18. Chiari, G., Giustetto, R., Druzik, J., Doehne, E., Ricchiardi, G.: Pre-columbian nanotechnology: reconciling the mysteries of the maya blue pigment. Appl. Phys. A **90**(1), 3–7 (2008)
19. Polette-Niewold, L.A., Manciu, F.S., Torres, B., Alvarado, M., Chianelli, R.R.: Organic/inorganic complex pigments: ancient colors Maya Blue. J. Inorg. Biochem. **101**(11), 1958–1973 (2007)
20. Sánchez del Rio, M., Martinetto, P., Reyes-Valerio, C., Dooryhee, E., Suárez, M.: Synthesis and acid resistance of Maya Blue pigment. Archaeometry **48**(1), 115–130 (2006)
21. Van Olphen, H.: Maya Blue: a clay-organic pigment? Science **154**(3749), 645–646 (1966)
22. Del Río, M.S., Martinetto, P., Somogyi, A., Reyes-Valerio, C., Dooryhee, E., Peltier, N., Alianelli, L., Moignard, B., Pichon, L., Calligaro, T., Dran, J.C.: Microanalysis study of archaeological mural samples containing Maya blue pigment. Spectrochim. Acta Part B **59**(10), 1619–1625 (2004)
23. Pacey, A.: Technology in world civilization: a thousand-year history. Philos. East West **43**(1), 155 (1993)
24. Verhoeven, J.D., Pendray, A.H., Dauksch, W.E.: The key role of impurities in ancient Damascus steel blades. JOM **50**(9), 58–64 (1998)
25. Reibold, M., Pätzke, N., Levin, A.A., Kochmann, W., Shakhverdova, I.P., Paufler, P., Meyer, D.C.: Structure of several historic blades at nanoscale. Cryst. Res. Technol. **44**(10), 1139–1146 (2009)
26. Srinivasan, S., Ranganathan, S.: India's Legendary Wootz steel: an advanced material of the ancient world. Iron & steel heritage of India Bangalore: National Institute of Advanced Studies, 69–82 (2004)
27. Levin, A.A., Meyer, D.C., Reibold, M., Kochmann, W., Pätzke, N., Paufler, P.: Microstructure of a genuine Damascus sabre. Cryst. Res. Technol. **40**(9), 905–916 (2005)
28. Pal, D., Sahu, C.K., Haldar, A.: Bhasma: the ancient Indian nanomedicine. J. Adv. Pharm. Technol. Res. **5**(1), 4 (2014)
29. Chaudhary, A.: Ayurvedic Bhasma: nanomedicine of ancient India—its global contemporary perspective. J. Biomed. Nanotechnol. **7**(1), 68–69 (2011)

30. Sarkar, P.K., Das, S., Prajapati, P.K.: Ancient concept of metal pharmacology based on Ayurvedic literature. Ancient Sci. Life **29**(4), 1 (2010)
31. Kumar, A.: A review of traditional anticancer nano-medicine: triphala. Pharma Innov. **1**, 3(7, Part B) (2016)
32. Kumar Pal, S.: The ayurvedic Bhasma: the ancient science of nanomedicine. Recent Pat. Nanomed. **5**(1), 12–18 (2015)
33. Paul, S., Chugh, A.: Assessing the role of Ayurvedic 'bhasms' as ethno-nanomedicine in the metal based nanomedicine patent regime. J. Intellect Prop. Rights **16**, 509–515 (2011)
34. Sharma, R., Prajapati, P.K.: Nanotechnology in medicine: leads from Ayurveda. J. Pharm. Bioallied Sci. **8**(1), 80 (2016)
35. Kapoor, R.C.: Some observations on the metal-based preparations in the Indian system of medicine. Indian J. Tradit. Knowl. **9**, 562–575 (2010)

Nanodeliverables

Arunkumar Palaniappan and Indulekha Singaravelu

Abstract Nanodimensional materials are currently being actively researched in the area of biomedical sciences for the development of better therapies, imaging modalities, and in some cases combination of both known as theranostics. In case of therapeutic applications, these nanoparticulate systems are now extensively researched in the area of targeted and controlled delivery of small molecule drugs and biological entities like genes, peptides, proteins, vaccines. Targeted delivery of these molecules using nanoparticles helps in achieving higher therapeutic concentrations in the diseased region, thus reducing the toxic effects to the surrounding healthier tissues. Based on the source materials, these nanoparticulate systems can be classified as (a) lipid nanomaterials, (b) polymeric nanomaterials, (c) inorganic nanomaterials, (d) carbon nanomaterials, (e) biomimetic nanomaterials, and (f) peptide nanostructures. In this chapter, these nanomaterial systems and their recent applications will be explained in much greater detail. In case of lipid nanomaterials, liposomes, solid lipid nanoparticles (SLNs) will be described in greater detail. Polymeric nanomaterials can be derived from naturally occurring polymers like chitosan, alginate, collagen, gelatin or can be derived from synthetic polymers like poly lactic-co-glycolic acid (PLGA), poly caprolactone (PCL), poly-L-lactic acid (PLLA). Stimuli-responsive nanoparticles which are obtained from stimuli-responsive polymeric materials are used primarily as on-demand drug delivery systems. Inorganic nanomaterials mainly consist of gold nanoparticles, silver nanoparticles, platinum nanostructures, copper nanoparticles, and other metallic oxide nanomaterials, which are widely used, in theranostic applications. Carbon nanomaterials mainly include graphene oxide, carbon nanotubes, fullerenes, carbon nanospheres, and nanodiamonds and are mainly explored as drug delivery agents for theranostic applications. Biomimetic nanoma-

A. Palaniappan
Division of Cardiovascular Diseases, Department of Internal Medicine,
University of Cincinnati, Cincinnati, OH 45267, USA
e-mail: palaniar@ucmail.uc.edu

I. Singaravelu (✉)
Division of Pharmaceutical Sciences,
James L. Winkle College of Pharmacy, University of Cincinnati, Cincinnati,
OH 45267, USA
e-mail: singaria@ucmail.uc.edu

M. Khoobchandani and A. Saxena (eds.), *Biotechnology Products in Everyday Life*,
EcoProduction, https://doi.org/10.1007/978-3-319-92399-4_2

terials are those which mimic the nature and properties of the biological components/constituents inside the body. Various cells such as erythrocytes, leukocytes, neutrophils, stem cells, immune T cells, and cancer cells are widely researched in transforming them as nanovesicles containing the therapeutic and/or imaging molecules to deliver at their target regions. Peptide nanomaterials are self-assembled nanostructures, which are currently explored in protein/drug delivery applications and in tissue engineering applications.

Keywords Nanomedicine · Drug delivery · Liposomes · Polymers · Inorganic Metal · Carbon · Biomimetic · Peptides · Self-assembly

1 Introduction

Drug delivery systems include technologies/systems, formulations, or sometimes even approaches to improve the therapeutic and pharmacological effects of therapeutic molecules, which can be small molecule drugs or other biopharmaceuticals like proteins, peptides, genes, vaccines [1]. Nanotechnology-based drug delivery systems are nanosized carriers (typically 1–100 nm) which have inherently better pharmacokinctic properties because of their small size (nanometer size), high surface to volume ratio [1]. Also, because of their large surface to volume ratio, a large number of targeting ligands can be bound on the surface of these nanoparticles, which aids in their precise disease targeting [1, 2].

1.1 Advantages of Nanodrug Delivery Systems

Nanodrug delivery systems have several advantages when compared to direct administration of therapeutic molecules. They are as follows: These nanocarriers can prevent degradation of therapeutic molecules in the circulation; they can be targeted to the desired location, and thereby they may have reduced toxicity to healthier cells/tissues; nanocarriers can be tuned to have prolonged circulation time by modifying their surface properties; they can load more amount of therapeutic molecules and can have controlled release property [1–4]. In case of cancer and other inflammation-related disorder therapy, because of the enhanced permeation and retention property, nanocarriers accumulate more in the solid tumor/other diseased region rather than in the normal tissue. This mechanism of increased accumulation of nanocarriers in the diseased region using the desired properties of the nanocarriers and the disease pathology (like enhanced permeation and retention (EPR) effect) is known as passive targeting [5]. On the other hand, active targeting of nanocarriers includes using specific ligands like peptides, proteins, antibodies, folates, and other disease-relevant biological moieties on the nanocarriers' surface that will be recognized by cells present in the diseased region [3, 5].

1.2 Classification of Nanodrug Delivery Systems

These nanocarriers can be classified into the following categories based on the source materials with which they are synthesized. They are as follows:

1. Lipid-based nanocarriers
2. Polymeric nanoparticles
3. Inorganic nanoparticles
4. Carbon nanomaterials
5. Biomimetic nanostructures
6. Peptide nanostructures.

2 Lipid-Based Nanodeliverables

2.1 Liposomes

Liposomes are self-assembled vesicular structure which has an aqueous core surrounded by bilayered phospholipids shell (Fig. 1). Liposomes are first synthesized by A. B. Bangham in 1961. Liposomes can typically be used to load both hydrophobic and hydrophilic drugs because of their amphiphilic nature. Hydrophilic drugs can be loaded in the hydrophilic aqueous core, while hydrophobic drugs can be loaded or impregnated within the lipophilic lipid bilayers [6]. Liposomes, because of their favorable physicochemical and biocompatible properties, are the most widely commercialized nanodrug delivery systems. There are a number of liposome-based nanoformulations available in the market, and a few are in clinical trials. Doxil, a PEGylated liposomal formulation for anticancer drug Doxorubicin, is the first commercialized liposomal drug formulation. PEGylation of liposomal formulation is done to prolong the circulation time of the liposomes and thereby enhance the passive tumor targeting using enhanced retention and permeation phenomenon in the solid tumor regions [6].

2.2 Stimuli-Responsive Liposomes

Thermo-sensitive liposomes, which are made of thermo-sensitive lipids, are also currently explored as an option for triggered drug release system for the treatment of cancer. In thermo-sensitive liposomes, the structure of lipid bilayer changes from a solid-gel phase to a liquid-crystalline phase at the melting phase transition temperature (T_m). Upon providing heat, this structural change is utilized as a drug release mechanism to release the encapsulated drug. DPPC (1,2-dipalmitoyl-sn-glycero-3-phosphocholine) because of its physiologically relevant T_m (~41.4 °C) is the

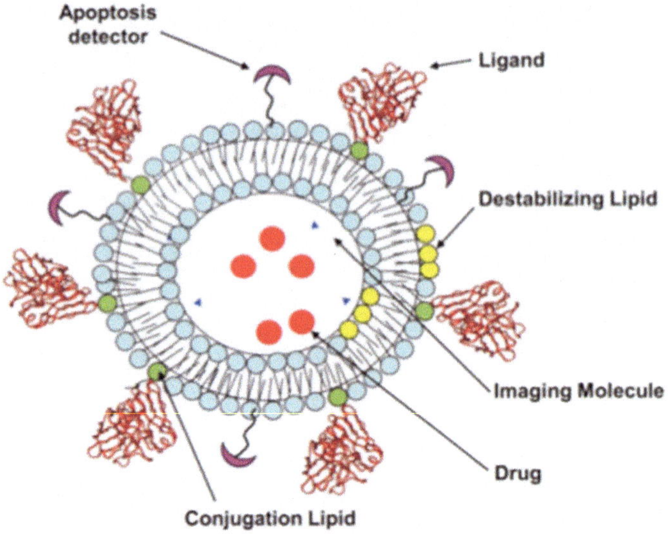

Fig. 1 An ideal liposomal drug delivery system for targeted and triggered drug delivery. Liposomes consist of bilayer made of phospholipid (cyan), a destabilizing (pore forming) phospholipid (yellow), conjugation lipid (green), targeting ligand attached to the liposome by conjugation lipid (brown), and a cell death marker such as an apoptotic detector (pink). The liposome is loaded with a chemotherapeutic agent (red) and an imaging agent (blue) in its aqueous core [6]

most predominantly used lipids for making thermo-sensitive liposomes. DSPC (1,2-distearoyl-sn-glycero-3-phosphocholine) having T_m of ~54.9 °C is often combined with DPPC in order to raise the T_m of the thermo-sensitive liposomes to hyperthermia conditions and prevent leakage of drug in undesired region [7].

The pH-sensitive liposomes remain intact in the circulation, are endocytosed in their intact form, and release their content inside the cytoplasm, which is used in the intracellular delivery of therapeutic molecules like drugs, nucleic acids. Typically, pH-sensitive liposomes are synthesized using pH-responsive lipids such as phosphatidylethanolamine (PE), cholesteryl hemisuccinate (CHEMS) and sometimes using polymers like poly(organophosphazenes), terminally alkylated copolymer of poly(N-isopropylacrylamide) (NIPAAm), and methacrylic acid-incorporated liposomes [8, 9].

Redox-sensitive liposomes are those liposomes, which release their contents intracellularly in response to the reducing environments inside the cytoplasm of cells like the presence of glutathione. These liposomes are mainly constituted by disulfide (–S–S–) bond containing lipids. This disulfide bond that stabilizes the lipid bilayer is cleaved upon exposure to the reducing environment inside the cells. Few of the examples of redox-sensitive liposomes include thiocholesterol-based cationic liposomes for DNA delivery, targeted delivery of Doxorubicin to human B lymphoma cells using disulfide linkage (mPEG-S-S-DSPE) [8, 9].

2.3 *Solid Lipid Nanoparticles and Nanostructured Lipid Carriers*

Though liposomes are most widely researched and commercialized lipid-based nanodrug delivery system, it suffers from the following drawbacks: short shelf life, poor stability, low drug encapsulation efficacy, rapid elimination by reticuloendothelial system (RES), cellular interactions or adsorption, and intermembrane transfer [6]. In order to overcome these drawbacks, solid lipid nanoparticles (SLNs) and nanostructured lipid carriers (NLCs) are introduced.

SLNs are spherical lipid nanostructures of size in the range from 40 to 1000 nm in diameter. SLNs are made using solid lipids in an aqueous solution in the presence of surfactants. The lipids used in preparation for SLN are fatty acids, steroids, waxes, monoglycerides, diglycerides, and triglycerides. These nanoparticles can be used in the encapsulation of both hydrophilic and hydrophobic drugs. Some of the key benefits of SLN other than the inherent biocompatibiltiy of lipids are as follows: ease of preparation, excellent physical stability, good drug release profiles, low cost of production, can be synthesized with little or no involvement of organic solvents. On the other hand, SLN has a few inherent disadvantages like the tendency to induce gelation, growth of lipid particles, and dynamics of polymorphic transitions and their low drug encapsulation because of the crystalline structure of solid lipids. In order to improve upon these properties of SLN, a new type of lipid nanostructures called NLC is introduced. In NLC, lipid phase consists of a mixture of both solid and liquid phase lipids, which results in a formless matrix that has a good stability and drug encapsulation efficiency. NLC is typically in the size range from 100 to 500 nm. The main method of the NLC synthesis is through high-pressure homogenization process. The other less frequently used methods involve microemulsification and solvent displacement. Some of the benefits of NLC when compared to SLN are as follows: good drug-loading capacity and minimized drug expulsion during storage (Fig. 2) [10].

3 Polymeric Nanocarriers

Polymeric nanodrug delivery systems have revolutionized the transport of therapeutic drugs and biomolecules inside the body in order to achieve effective delivery and enhanced therapeutic efficiency. The main challenge is to attain the maximum therapeutic index with lower dose of drug and also to reach the target site by evading the elimination mechanism in the body [11]. However, it is very critical to understand the carrier properties such as improved bioavailability and pharmacokinetics of the drugs, especially for the water-insoluble drugs and methods in formulating the drug-loaded carriers. Understanding the size effects and the biological properties, i.e., nano–bio interface, is essential in developing the nanocarriers for delivery applications. The physicochemical properties of the polymers, which include their surface

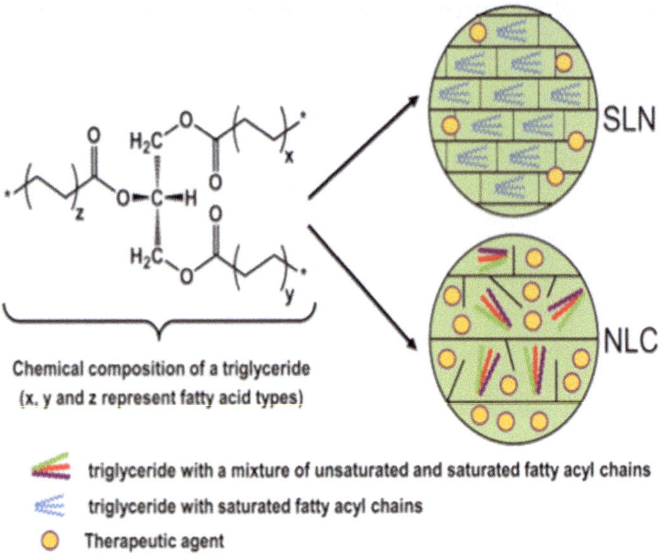

Fig. 2 Schematic representation of solid lipid nanoparticles (SLN) and nanostructured lipid carriers (NLC) [6]

charge, functionalities, biodegradability, biocompatibility, transport mechanism, and drug release mechanism, are the most important parameters to be well understood in their role as nanodeliverables [1]. Various polymers are employed in preparation for nanoparticles (NPs). Based on their origin, they are classified as naturally derived polymers and synthetic polymers.

3.1 Synthetic Polymers

Initially, polymeric NPs are synthesized using synthetic polymers that were usually non-degradable like poly (methyl methacrylate) (PMMA), polyacrylamide, polystyrene, and polyacrylates [12]. These polymers are designed in such a way that they are eliminated through urine or feces quickly after they deliver their cargo inside the body. Furthermore, they are used extensively in wound healing [13], antimicrobial delivery [14] where their role has been needed for long-term use. The synthesis of these polymers and polymeric NPs is simple and is easy to control the parameters/factors involved in their nanoformulation reactions. However, serious toxicity and inflammatory issues are reported at most times with these non-degradable polymers. Hence, biodegradable polymers came into picture and are employed widely in the drug delivery applications. Few biodegradable synthetic polymers include and are not limited to the following: poly (lactic-co-glycolic acid), (PLGA), polycaprolactone (PCL) which are biodegradable and are widely used in various therapeutics delivery applications [15–17].

3.2 *Natural Polymers*

Natural polymers are the biopolymers derived from biological materials and are made up of polysaccharides, protein, or any other biomacromolecules. Polymers like chitosan, chitin, dextran, cyclodextrins, alginate, pectin, xanthan gum and proteins like albumin, gelatin are some of the few naturally derived polymeric structures and are FDA-approved polymers for biomedical applications. These polymers play a significant role in drug delivery as nanocarriers with many advantages over the synthetic polymers. They are as follows: Natural polymers are inexpensive source of materials that are abundantly found in nature, and NPs of these polymers are enriched with the presence of more reactive sites for bioconjugation like attaching drugs, proteins, and targeting ligands. They are biocompatible and biodegradable, hence rendering them more attention in drug delivery applications [18]. Being natural biopolymers, they are mostly hydrophilic in nature, non-toxic, and highly stable in the biological fluids. Some of these polymeric nanostructures form non-covalent bonds with the biological tissues due to their muco-adhesive properties, and hence they are called muco-adhesive polymers. They are most explored in the delivery of chemotherapeutics for the cancer therapy, gene delivery like siRNA, and also in the delivery of proteins and growth factors in tissue engineering applications and regenerative medicine [19].

3.3 *Stimuli-Responsive Polymers (SRPs)-Based Nanocarriers*

Stimuli-responsive polymers are those that are potentially designed to sense and act in response to the stimuli in their environment. SRP-based nanocarriers are fabricated in such a way to "respond to their stimuli" and thereby release the drug "on demand". They are new generation of "smart" polymeric NPs that respond to the stimuli in the microenvironment. The stimuli can be extrinsically applied or intrinsically available in the physiological environment. External triggers can be heat, light, electric and magnetic field, whereas intrinsic triggers can be cellular pH-shift, redox, ionic microenvironment of the specific tissues, and enzyme over-expression in certain pathological states, host–guest recognitions, and antigen–antibody interactions [20].

Owing to this smart behavior of SRP nanocarriers, they are potential delivery systems to deliver drugs, dual delivery of contrast agents, and chemotherapeutics in cancer theranostic applications in response to the stimuli in the microenvironment of the cancer tissue. Furthermore, they are widely used in targeted delivery of the cells or growth factors or genes or proteins like insulin to the specific tissues for various

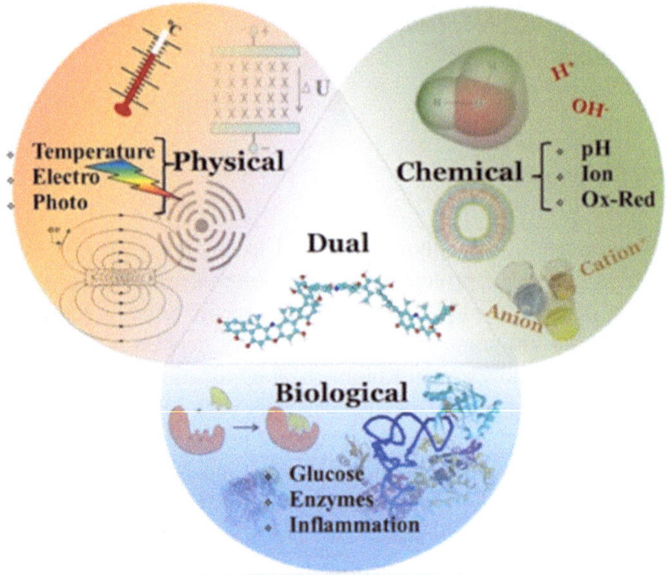

Fig. 3 Various stimuli for responsive polymeric system [25]

applications [21]. The nanocarriers are tailor-made into dual responsive or multi-responsive by fabricating them with one or more polymers that responds to more than one stimulus. For example, dual-responsive nanocarriers can be temperature/pH or magnetic/pH or pH/redox, or pH/enzyme (Fig. 3).

Thermo-responsive polymers such as poly(N-isopropylacrylamide) (PNIPAAM) and poly (N vinyl caprolactam) are most widely used in drug delivery applications [22, 23]. Similarly, chitosan, poly(acrylic acid) (PAA), poly(methacrylic acid) (PMAA) are pH-responsive polymers and are mostly used in targeting intracellular delivery like lysosomes (low pH) and in cancer (due to low pH in tumor environment) [24]. Hence, SRP nanocarriers have the potential role in on-demand and triggered delivery for various diagnostic and therapeutic applications.

4 Inorganic Nanoparticles

Inorganic NPs have shown a prominent progress in the development of nanomedicines by their outstanding material properties. Metal and semiconductor NPs have proved to exhibit excellent therapeutic and imaging properties with high molar extinction coefficient, size/shape-dependent tunable optical properties, and high stability against photodegradation [26–28]. Inorganic NPs are broadly comprised of semiconductors (quantum dots), silica-based, metals, metal oxides, lanthanum-doped NPs (Fig. 4). Their size, surface charge, and the surface complex-

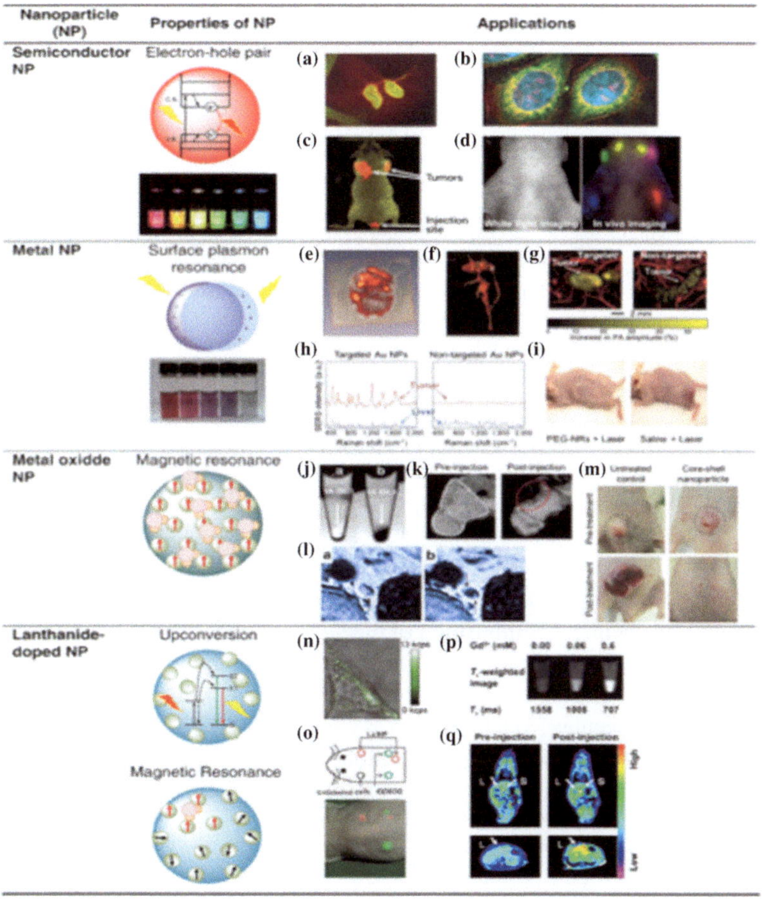

ity enable them to conjugate the drug molecules and the targeting ligands and achieve targeted delivery of therapeutics to target site inside the body with high specificity. Surface engineering of such nanostructures allows them to exhibit exceptional material properties with biological interactions, which can measure/detect the changes or abnormalities at the cellular level.

4.1 *Quantum Dots (QDs)*

Semiconductor or QDs are the fluorescent nanocrystals with size ranging typically from 1 to 10 nm. Their size is very critical and is responsible for the quantum properties such as electronic structure and discrete energy levels resulting in novel optical

◀**Fig. 4** Various inorganic nanoparticles and its applications. **a** Fluorescence imaging of mouse 3T3 fibroblast cells using quantum dots (QDs). **b** Multiplexed fluorescence image depicting five-color QD staining of fixed human epithelial cells. **c** Fluorescence imaging for detecting human prostate cancer with targeted QDs. **d** Multiplexed fluorescence imaging of five different lymph nodes using five different QDs. **e** Three-dimensional (3-D) OCT view of sentinel lymph nodes morphology with poly(ethyleneglycol)-coated gold nanorods (PEGylated Nurs). **f** Three-dimensional CT angiogram image of the heart and great vessels with PEGylated gold NPs. **g** Photoacoustic imaging for detection of human melanoma with targeted gold nanocages. **h** Surface-enhanced Raman spectra for the detection of human squamous cell carcinoma using targeted gold NPs. **i** Photothermal therapy of tumor-bearing mice using PEGylated gold NRs. **j** Magnetic resonance (MR) detection of human breast cancer cells without (**a**) and with (**b**) targeted iron oxide NPs. **k** MR imaging of human colon tumor using targeted iron oxide NPs [108]. (l) MR images of a metastatic lymph node before (**a**) and after (**b**) administration of iron oxide NPs [109]. **m** Magnetic hyperthermia treatments of tumor-bearing mice using superparamagnetic $CoFe_2O_4@MnFe_2O_4$ NP (core–shell NP). **n** Upconverted luminescence imaging of NIH 3T3 murine fibroblast cells with $NaYF_4:Yb^{3+}/Er^{3+}$ NPs. **o** In vivo multiplexed photoluminescence imaging of mouse obtained by simultaneously using $NaYF_4:Yb^{3+}/Tm^{3+}$ NP (La NP) and QD (QD800). **p** T1-weighted MR images of human breast cancer cells with $NaGdF_4:Yb^{3+}/Er^{3+}$ NPs. **q** Color-mapped MR images of a mouse with $Tm^{3+}/Er^{3+}/Yb^{3+}$ co-doped $NaGdF_4$ NPs [29]

and electronic properties, as they are semiconducting materials [30]. Tuning the size, shape, and chemical composition of the QDs can alter the optical emission and absorption. Hence, QDs render a potential role in a wide range of imaging applications, cell labeling, and single molecule tracking than other fluorescent/luminescent dyes and small molecules [31]. QDs such as CdSe, CdS, CdTe [32] have been extensively studied for the bio-imaging applications because of their near-infrared (NIR) deep tissue penetration capability. However, being heavy metals, they are found to be toxic in vivo and thus ZnTe/ZnSe QDs emerged with better fluorescent emission property and low toxicity [33].

4.2 Metal and Metal Oxide Nanoparticles

Among diverse physical properties of the metal NPs, two distinct properties such as superparamagnetism and surface plasmon resonance (SPR) are essential for biomedical applications. SPR is the phenomenon of oscillation of the free electrons at the interface of the metal NPs and surrounding medium resulting in the resonance with the electromagnetic field [27]. These NPs absorb and scatter the incident light upon the excitation of SPR oscillations; thus, these scattered lights from the NPs can be used for biological imaging of dark-field optical microscopy and optical coherence tomography (OCT) [34]. Their large absorption has allowed for the photoacoustic tomography (PA) [69 of 10] and photothermal cancer therapy [35]. They also possess large absorption coefficients that render X-ray imaging, two-photon luminescence (T-PL), contrast-based imaging computed tomography (CT), and surface-enhanced Raman scattering (SERS) [36]. Noble metals like gold (Au) are more extensively studied in various forms of nanostructures like Au nanoshells, Au nanorods, Au nanocages, Au clusters, and Au nanowires with different optical and physical properties. Similarly, silver (Ag) and copper (Cu) nanoparticles have exhibited strong antibacterial effects against various pathogenic and resistant bacterial strains resulting in high impact in the wound healing and antimicrobial applications [29].

Superparamagnetic property is mostly exhibited by the metal oxide nanoparticles with size range 10–20 nm such as iron oxide (Fe_2O_3 and Fe_3O_4) with wide applications in magnetic resonance imaging (MRI) and magnetic field-induced hyperthermia therapy predominantly in cancer [37]. TiO_2 NPs are used as photocatalysts in photodynamic therapy (PDT). Moreover, paramagnetic ions like gadolinium, ytterbium lanthanum, and europium have a potential role in MRI applications [29].

4.3 Silica-Based Nanoparticles

Amorphous silica-based NPs possess the inherent non-toxic, biocompatible property in the biological tissues for the targeted delivery of therapeutics and also for the diagnostic purpose. Mesoporous silica NPs have high loading efficiency of drugs, and small biomolecules and their flexibility in the surface properties for labeling of cells and peptides have also been studied. Labeled with targeting ligands and proteins, they are extensively used in diagnostic and tracking applications [38].

5 Carbon Nanomaterials

5.1 Fullerene

It is a molecule of carbon and exists in the form of a hollow sphere, ellipsoid, or tube. Spherical fullerene is called as buckyball or buckminsterfullerene (C60). These are water insoluble and form aggregates in aqueous solutions. In order to improve the solubility of fullerene in aqueous solutions, functionalization with hydrophilic biocompatible materials is explored. Few of the materials studied for functionalization include amino acids, carboxylic acids, polyhydroxyl groups, amphiphilic polymers. In addition to its application as a drug and gene delivery systems, fullerene is also explored as photosensitive agents, diagnostic agents in the form of endofullerenes/metallofullerenes, antiviral molecule, and antioxidant material [39, 40].

5.2 Graphene

Over the past 40 years, carbon nanomaterials are one of the most fascinating nanomaterials which have attracted much attention among scientific communities because of

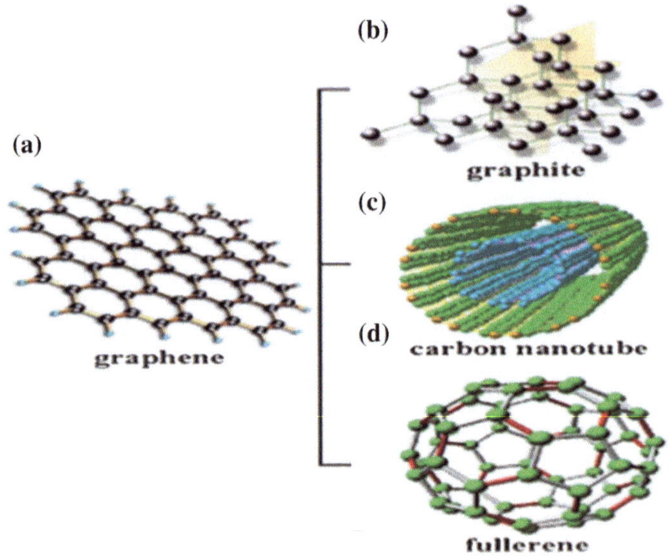

Fig. 5 Schematic representation of the four allotropic forms of sp^2 carbon: graphene (**a**), graphite (**b**), nanotubes (**c**), and fullerene (**d**) [43]

their unique electrical, thermal, mechanical, and other physicochemical properties [41]. Graphene is a single-atom-thick layer of sp^{2-} hybridized carbon atoms arranged in a honeycomb lattice. It is the basic building block of other carbon allotropes including 0-D fullerenes, 1-D carbon nanotubes, and 3-D graphite [42] (Fig. 5). In addition to their excellent electrochemical and optical properties, graphene can adsorb a variety of aromatic biomolecules through a π–π stacking interaction and/or electrostatic interaction, which make them ideal materials for biosensors and drug delivery applications [42].

5.3 Graphene Oxide and Reduced Graphene Oxide

Graphene-based nanosheets include graphene nanosheets, graphene oxide (GO) nanosheets, and reduced graphene oxide (rGO) nanosheets. The two-dimensional structure of graphene nanosheets provides excellent surface area for the drugs and other therapeutic molecules to get loaded and released in a controlled manner. When compared to graphene nanosheets, GO and rGO nanosheets are studied more in terms of drug delivery applications because of their favorable physicochemical properties. In case of GO nanosheets, because of the presence of hydroxyl (–OH), epoxy (–O–) and carboxyl (–COOH) group, electronegative atoms of drug molecules can interact with it through hydrogen bonding. Moreover, the ionizable carboxyl group at the edge of GO allows electrostatic interactions with drug molecules. Also, on the basal

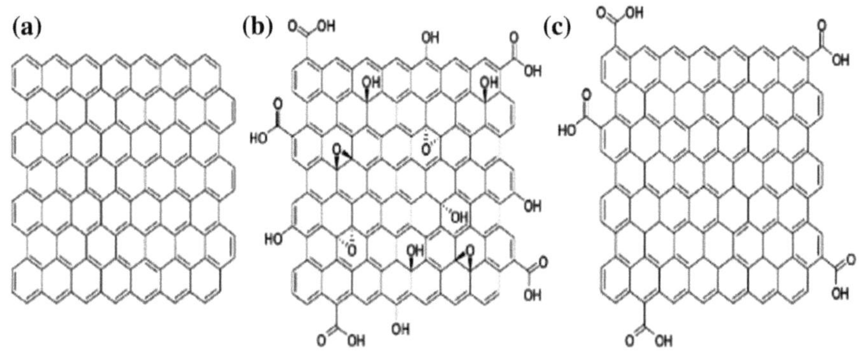

Fig. 6 Structures of graphene-based nanosheets. Structures of graphene (**a**), GO (**b**), and rGO (**c**) [42]

planar region, drugs can interact through hydrophobic interaction as well as π–π stacking interaction (Fig. 6).

The rGO nanosheets have fewer oxygen-containing functional groups and thus more hydrophobic when compared to GO. The interaction of drug with rGO is mainly due to hydrophobic interaction and π–π stacking interaction because of their hydrophobic nature.

These graphene nanosheets are used in the delivery of chemotherapeutic drugs like Doxorubicin, Paclitaxel, Fluorouracil, Camptothecin, Methotrexate, platinum complexes. These nanosheets are also used in the delivery of photosensitizers along with chemotherapeutics, biologic drugs like nucleic acid, proteins, and peptides. On the other hand, these nanosheets suffer from toxicity issues in in vivo conditions and functionalization of the nanosheet surface using biocompatible materials is found to have better biocompatibility [44]. Some of the most widely researched materials used for the functionalization include and are not limited to the following: polyethylene glycol (PEG), poloxamer 407, poly (amido amine), chitosan, gelatin, bovine serum albumin, low molecular weight heparin, phospholipids [42, 44, 45].

5.4 Carbon Nanotubes

Carbon nanotubes (CNTs) are first discovered by Sumio Iijima in the year 1991. CNTs are allotropes of carbon, which are formed when graphene sheets are rolled and held together by van der Waals interactions. CNT can be classified as single-walled CNT (SWCNT) and multi-walled CNT (MWCNT) based on the number of graphene layers involved in the formation of CNT [41]. SWCNT has a diameter between 0.4 and 2 nm, while a MWCNT has diameter between 2 and 100 nm. Having a typical length of a few microns, CNTs have an aspect ratio (L/D) of 1:1000. CNT has excellent properties like high tensile strength, excellent chemical and thermal stability, a large surface area which can be used to conjugate therapeutic molecules like drugs, proteins, peptides, and nucleic acids for drug delivery applications. The internalization of CNT inside cells is found to occur due to the following mechanisms: endocytosis, phagocytosis, or passive diffusion. Energy-dependent endocytosis and phagocytosis are the key mechanisms by which SWCNT are engulfed by cells, while MWCNT is taken in by passive diffusion [44].

CNTs readily form aggregates in the physiological solutions, which necessitates the need for the functionalization of CNT for biomedical applications. The functionalization techniques include the following: (a) covalent functionalization involving grafting of molecules on the sp^2 carbon atoms of the π-conjugated skeleton of the CNT, and these molecules can be hydrophilic polymers like polyethylene glycol (PEG); (b) non-covalent functionalization using amphiphilic surfactant molecules or polymers; and (c) click chemistry-mediated functionalization reactions [46]. CNTs are currently explored as a drug delivery agent for the delivery of anticancer drugs, nucleic acid therapy, and also as a photothermal agent for the treatment of cancer [42–46].

5.5 *Carbon Nanospheres*

Carbon nanospheres are formed as a result of pairing of pentagonal and heptagonal carbon rings which results in spherical arrangements. In case of carbon nanospheres, the graphite sheets are not closed shells but rather waving flakes that follow the curvature of the sphere, creating many open edges at the surface. These provide reactive dangling bonds on the surface, which enhances adsorption of molecules interacting with carbon nanospheres. This property in addition to large surface area of carbon nanospheres is used in the loading of therapeutic molecules for drug delivery applications. In addition to these, carbon nanospheres also have intrinsic fluorescence property which could be used for intracellular drug delivery and its subsequent tracking applications [47, 48].

5.6 *Nanodiamond*

Nanodiamond is first produced by detonation in USSR in the 1960s. Nanodiamonds are found to be less toxic than other carbon nanomaterials and thus are best suited for biomedical applications like drug delivery and biomedical imaging. Typically, nanodiamonds are of size 4–5 nm, but because of the aggregation issues they tend to be larger aggregate particles. Because of the presence of dangling bonds on the surface, therapeutic molecules can be bound on the particles' surface for drug delivery applications. Nanodiamonds are also found to have intrinsic fluorescent properties and by conjugation with fluorophores on its surface are explored as a potential biomedical-imaging agent. Nanodiamond is currently used in delivery of Doxorubicin to treat drug-resistant breast cancer (4T1) and liver cancer (LT2-M) models. In addition to this, nanodiamonds coated with polyethylenimine 800 (PEI800) were studied for nucleic acids delivery. Because of their surface property, nanodiamonds are also explored as a protein mimics which can mimic enzymatic functions [49].

6 Polymer–Inorganic Composite Nanostructures

Polymer–inorganic composite NPs have revolutionized a wide range of applications with additive and enhanced physicochemical/material properties, respectively, when they interact with the biological system. Various factors such as biocompatibility, drug-loading efficiency, drug release kinetics, optical and imaging properties were taken into consideration in designing composite NPs to overcome the undesired effects of the inorganic and polymer NPs separately. Multifunctional composite polymer-coated inorganic NPs were designed as on-demand, targeted delivery systems to achieve the desired therapeutic effect with high efficacy [23, 50].

7 Biomimetic Nanostructures for Delivery Applications

Recently, the leveraging of the targeting mechanism in drug delivery has occurred where the specific targeting moieties have been mimicked by the structures that are evolved from the biological interactions within the body [51]. In order to overcome certain biological barriers within the body by mimicking the nature and properties of the biological components/constituents inside the body, biomimetic nanostructures are fabricated. These biomimetic structures act as stealth carriers to evade the immune system of the body and deliver their cargo. Hence, two strategies or approaches were followed to obtain the desired nano-based delivery systems for therapy and imaging, i.e., bottom-up and top-down strategies. The "bottom-up" approach comprises of synthesizing nanocarriers such as polymeric/protein nanogels, liposomes, micelles, dendrimers, polymerosomes from the biocompatible and biodegradable materials. Furthermore, it includes the strategy of targeting the nanocarriers with antibodies, peptides, and other ligands to achieve the right destination of the particles inside the body. The positive aspect of this approach is the controllability in the physicochemical parameters in the nanoparticle synthesis, encapsulation, and also surface functionalization with ligands and molecules to bind the receptors of the target cells. However, they fail to reproduce the complexity of the cell membranes on the surface of the nanocarriers and hence remain incompetent in evading the mononuclear phagocytic system (MPS) [52].

The "top-down" approach encompasses the synthesis of nanocarriers derived from the biological constituents such as cells, bacteria, viruses, and their cell components in the form of nanovesicles (Fig. 7). The major advantage of this top-down approach is that these biomimetic nanostructures can easily permeate across the bio-

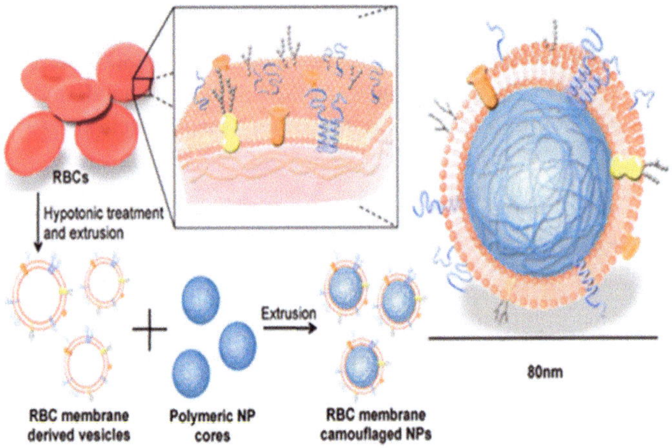

Fig. 7 Biomimetic nanostructures—synthesis of RBC membrane-coated NPs [54]

logical membrane and evade MPS within the body. However, the limitation of this approach is the size and homogeneity of the final formulation, which is uncertain [53]. Such biomimetic nanocarriers are synthesized from various cells using the top-down approach, namely erythrocytes, leukocytes, neutrophils, cancer cells, stem cells, dendritic cells (DC), virus-like proteins (VLPs), bacteria, and other eukaryotic cells.

Biomimetic nanostructures can be synthesized from various cells of origin in three ways: (1) entire cell as carrier, (2) extracellular membrane as exosomes or vesicles, and (3) as cell membrane-coated nanoparticles which are depicted in Fig. 7. They serve the purpose of delivering the cargo at the target site by retaining their biological functions and their targeting specificity without any alteration [55, 56]. For instance, RBCs are used as the biomimetic carrier in delivering the drug/biomolecules without altering their biological properties like evading the immunogenic reactions in the body and providing long circulating life for the carriers [50]. Similarly, leukocytes [57], platelets [58], stem cells [59], cancer cells [60] are employed in various strategies to achieve the targeted delivery of drugs with low immunogenicity, extended long-term circulation, and hence enhanced bioavailability. The cells and cell membrane are engineered in such a way to achieve specificity in delivering at the target site by overcoming opsonization and RES clearance in the circulation [55].

In addition to these, pathogen-derived vectors are also extensively studied which includes bacterial ghosts and virus-like particles (VLPs). The bacterial membrane derived from gram-positive and gram-negative bacteria is used as carriers to deliver drugs/biomolecules in the form of membrane vesicles (MVs) as they are innate carriers of proteins, peptides, nucleic acids, and enzymes. They are more extensively used for vaccine delivery against infections and cancer-carrying gene targets to achieve gene silencing and other therapeutic applications [61]. VLPs are also designed for vaccine delivery against staphylococcus and other bacterial infections such as bacteriophage membrane loaded with antigenic nucleoproteins [62].

8 Self-assembled Peptide Nanostructures

Self-assembled peptide nanostructures are recent discovery in the field of biological nanomaterials. These are nanostructures made due to self-assembly of a few amino acids, which can be either natural or synthetic amino acids. These peptides can self-assemble into various shapes like spherical nanovesicles, nanoropes, nanotubes, micelles, nanofibers based on their amino acids sequence, and the peptide concentration in the solution (Fig. 8).

Fung et al. used various forms of a self-assembled peptide EAK16 for encapsulating and effectively delivering a hydrophobic anticancer drug Ellipticine in two different cancer cell lines [64]. Koutsopoulos et al. studied the possibility of encapsulating protein molecules in a self-assembled peptide nanofiber hydrogel scaffold and also proved the possibility of controlled release of these proteins [65]. The proteins used are lysozyme, trypsin inhibitor, BSA, and IgG. It was found that the protein release rate depended on proteins' molecular weight and peptide concentration in the formulation. These peptide nanostructures are also explored as 3D scaffold in tissue engineering and regenerative medicine [63, 66].

Conflicts of Interest The authors declare no conflict of interest.

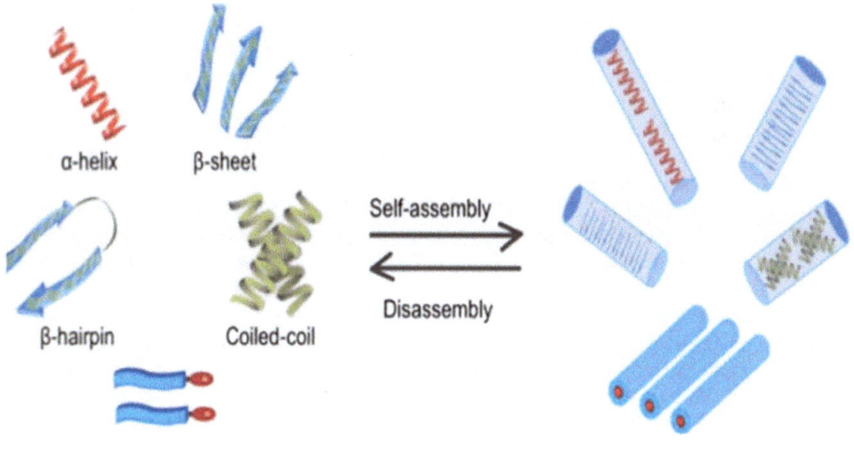

Fig. 8 Schematic illustration of the formation of peptide-based supramolecular hydrogels [63]

References

1. Kumar, A., Mansour, H.M., Friedman, A., Blough, E.R.: Nanomedicine in drug delivery. 469 (2013)
2. Devadasu, V.R., Bhardwaj, V., Kumar, M.N.V.R.: Can controversial nanotechnology promise drug delivery? Chem. Rev. **113**, 1686–1735 (2013). https://doi.org/10.1021/cr300047q
3. Parveen, S., Misra, R., Sahoo, S.K.: Nanoparticles: a boon to drug delivery, therapeutics, diagnostics and imaging. Nanomed. Nanotechnol. Biol. Med. **8**, 147–166 (2012). https://doi.org/10.1016/j.nano.2011.05.016
4. Wilczewska, A.Z., Niemirowicz, K., Markiewicz, K.H., Car, H.: Nanoparticles as drug delivery systems. Pharmacol Rep. **64**, 1020–1037 (2012). https://doi.org/10.1016/S1734-1140(12)70901-5
5. Nakamura, Y., Mochida, A., Choyke, P.L., Kobayashi, H.: Nanodrug delivery: is the enhanced permeability and retention effect sufficient for curing cancer? Bioconjug. Chem. **27**, 2225–2238 (2016). https://doi.org/10.1021/acs.bioconjchem.6b00437
6. Puri, A., Loomis, K., Smith, B., et al.: Lipid-based nanoparticles as pharmaceutical drug carriers: from concepts to clinic. Crit. Rev. Ther. Drug Carrier Syst. **26**, 523–580 (2009). doi: 7a0c3bee026fb778,313cd9ee34e9e406 [pii]
7. Kneidl, B., Peller, M., Winter, G., et al.: Thermosensitive liposomal drug delivery systems: state of the art review. Int. J. Nanomed. **9**, 4387–4398 (2014). https://doi.org/10.2147/IJN.S49297
8. Movahedi, F., Hu, R.G., Becker, D.L., Xu, C.: Stimuli-responsive liposomes for the delivery of nucleic acid therapeutics. Nanomed. Nanotechnol. Biol. Med. **11**, 1575–1584 (2015). https://doi.org/10.1016/j.nano.2015.03.006
9. Ganta, S., Devalapally, H., Shahiwala, A., Amiji, M.: A review of stimuli-responsive nanocarriers for drug and gene delivery. J. Control Release **126**, 187–204 (2008). https://doi.org/10.1016/j.jconrel.2007.12.017
10. Iqbal, M.A., Md, S., Sahni, J.K., et al.: Nanostructured lipid carriers system: recent advances in drug delivery. J. Drug Target. **20**, 813–830 (2012). https://doi.org/10.3109/1061186X.2012.716845
11. Crucho, C.I.C.: Stimuli-responsive polymeric nanoparticles for nanomedicine. ChemMedChem **10**, 24–38 (2015). https://doi.org/10.1002/cmdc.201402290
12. Turos, E., Shim, J.Y., Wang, Y., et al.: Antibiotic-conjugated polyacrylate nanoparticles: new opportunities for development of anti-MRSA agents. Bioorg. Med. Chem. Lett. **17**, 53–56 (2007). https://doi.org/10.1016/j.bmcl.2006.09.098
13. Greenhalgh, K., Turos, E.: In vivo studies of polyacrylate nanoparticle emulsions for topical and systemic applications. Nanomed. Nanotechnol. Biol. Med. **5**, 46–54 (2009). https://doi.org/10.1016/j.nano.2008.07.004
14. Maya, S., Indulekha, S., Sukhithasri, V., et al.: Efficacy of tetracycline encapsulated O-carboxymethyl chitosan nanoparticles against intracellular infections of *Staphylococcus aureus*. Int. J. Biol. Macromol. **51**, 392–399 (2012). https://doi.org/10.1016/j.ijbiomac.2012.06.009
15. Arunkumar, P., Indulekha, S., Vijayalakshmi, S., Srivastava, R.: Synthesis, characterizations, in vitro and in vivo evaluation of Etoricoxib-loaded Poly (Caprolactone) microparticles-a potential Intra-articular drug delivery system for the treatment of Osteoarthritis. J. Biomater. Sci. Polym. Ed. **27**, 303–316 (2016). https://doi.org/10.1080/09205063.2015.1125564
16. Arunkumar, P., Indulekha, S., Vijayalakshmi, S., Srivastava, R.: Poly (caprolactone) microparticles and chitosan thermogels based injectable formulation of etoricoxib for the potential treatment of osteoarthritis. Mater. Sci. Eng. C **61**, 534–544 (2016). https://doi.org/10.1016/j.msec.2015.12.039
17. Snima, K.S., Arunkumar, P., Jayakumar, R., Lakshmanan, V.K.: Silymarin encapsulated Poly(D, L-lactic-co-glycolic acid) nanoparticles: a prospective candidate for prostate cancer therapy. J. Biomed. Nanotechnol. **10**, 559–570 (2014). https://doi.org/10.1166/jbn.2014.1735

18. Hamid Akash, M.S., Rehman, K., Chen, S.: Natural and synthetic polymers as drug carriers for delivery of therapeutic proteins. Polym. Rev. **55**, 371–406 (2015). https://doi.org/10.1080/15583724.2014.995806
19. Banik, B.L., Fattahi, P., Brown, J.L.: Polymeric nanoparticles: the future of nanomedicine. Wiley Interdisc. Rev Nanomed. Nanobiotechnol. **8**, 271–299 (2016). https://doi.org/10.1002/wnan.1364
20. Chipman, S.D., Oldham, F.B., Pezzoni, G., Singer, J.W.: Biological and clinical characterization of paclitaxel poliglumex (PPX, CT-2103), a macromolecular polymer-drug conjugate. Int. J. Nanomed. **1**, 375–383 (2006). https://doi.org/10.2147/nano.2006.1.4.375
21. Yang, D., Van, S., Liu, J., et al.: Physicochemical properties and biocompatibility of a polymer-paclitaxel conjugate for cancer treatment. Int. J. Nanomed. **6**, 2557–2566 (2011). https://doi.org/10.2147/IJN.S25044
22. Indulekha, S., Arunkumar, P., Bahadur, D., Srivastava, R.: Thermoresponsive polymeric gel as an on-demand transdermal drug delivery system for pain management. Mater. Sci. Eng. C **62**, 113–122 (2016). https://doi.org/10.1016/j.msec.2016.01.021
23. Indulekha, S., Arunkumar, P., Bahadur, D., Srivastava, R.: Dual responsive magnetic composite nanogels for thermo-chemotherapy. Colloids Surf. B. Biointerfaces **155**, 304–313 (2017). https://doi.org/10.1016/j.colsurfb.2017.04.035
24. Mano, J.F.: Stimuli-responsive polymeric systems for biomedical applications. Adv. Eng. Mater. **10**, 515–527 (2008). https://doi.org/10.1002/adem.200700355
25. Cabane, E., Zhang, X., Langowska, K., et al.: Stimuli-responsive polymers and their applications in nanomedicine. Biointerphases **7**, 1–27 (2012). https://doi.org/10.1007/s13758-011-0009-3
26. Michalet, X.: Quantum dots for live cells, in vivo imaging, and diagnostics. Science **307**(80), 538–544 (2005). https://doi.org/10.1126/science.1104274
27. Jain, P.K., Huang, X., El-Sayed, I.H., El-Sayed, M.A.: Noble metals on the nanoscale: optical and photothermal properties and some applications in imaging, sensing, biology, and medicine. Acc. Chem. Res. **41**, 1578–1586 (2008). https://doi.org/10.1021/ar7002804
28. Biju, V., Itoh, T., Anas, A., et al.: Semiconductor quantum dots and metal nanoparticles: syntheses, optical properties, and biological applications. Anal. Bioanal. Chem. **391**, 2469–2495 (2008). https://doi.org/10.1007/s00216-008-2185-7
29. Nam, J., Won, N., Bang, J., et al.: Surface engineering of inorganic nanoparticles for imaging and therapy. Adv. Drug Deliv. Rev. **65**, 622–648 (2013). https://doi.org/10.1016/j.addr.2012.08.015
30. Bawendi, M.: The quantum mechanics of larger semiconductor clusters. Annu. Rev. Phys. Chem. **41**, 477–496 (1990). https://doi.org/10.1146/annurev.physchem.41.1.477
31. Hussain, S., Won, N., Nam, J., et al.: One-pot fabrication of high-quality InP/ZnS (Core/Shell) quantum dots and their application to cellular imaging. ChemPhysChem **10**, 1466–1470 (2009). https://doi.org/10.1002/cphc.200900159
32. Peng, Z.A., Peng, X.: Formation of high-quality CdTe, CdSe, and CdS nanocrystals using CdO as precursor [6]. J. Am. Chem. Soc. **123**, 183–184 (2001). https://doi.org/10.1021/ja003633m
33. Bang, J., Park, J., Lee, J.H., et al.: ZnTe/ZnSe (core/shell) type-II quantum dots: Their optical and photovoltaic properties. Chem. Mater. **22**, 233–240 (2010). https://doi.org/10.1021/cm9027995
34. Chen, J., Saeki, F., Wiley, B.J., et al.: Gold nanocages: bioconjugation and their potential use as optical imaging contrast agents. Nano Lett. **5**, 473–477 (2005). https://doi.org/10.1021/nl047950t
35. Hirsch, L.R., Stafford, R.J., Bankson, J.A., et al.: Nanoshell-mediated near-infrared thermal therapy of tumors under magnetic resonance guidance. Proc. Natl. Acad. Sci. **100**, 13549–13554 (2003). https://doi.org/10.1073/pnas.2232479100
36. Qian, X., Peng, X.-H., Ansari, D.O., et al.: In vivo tumor targeting and spectroscopic detection with surface-enhanced Raman nanoparticle tags. Nat. Biotechnol. **26**, 83–90 (2007). https://doi.org/10.1038/nbt1377

37. Chen, M., Nikles, D.E.: Synthesis, self-assembly, and magnetic properties of Fe x Co y Pt 100-x - y nanoparticles. Nano Lett. **2**, 211–214 (2002). https://doi.org/10.1021/nl015649w
38. Heneweer, C., Gendy, S.E., Peñate-Medina, O.: Liposomes and inorganic nanoparticles for drug delivery and cancer imaging. Ther. Deliv. **3**, 645–656 (2012). https://doi.org/10.4155/tde.12.38
39. Bakry, R., Vallant, R.M., Najam-ul-Haq, M., et al.: Medicinal applications of fullerenes. Int. J. Nanomed. **2**, 639–649 (2007). https://doi.org/10.1016/S0081-1947(08)60578-0
40. Montellano, A., Da Ros, T., Bianco, A., Prato, M.: Fullerene C60 as a multifunctional system for drug and gene delivery. Nanoscale **3**, 4035 (2011). https://doi.org/10.1039/c1nr10783f
41. Liu, Z., Robinson, J.T., Tabakman, S.M., et al.: Carbon materials for drug delivery & cancer therapy. Mater. Today **14**, 316–323 (2011). https://doi.org/10.1016/S1369-7021(11)70161-4
42. Shim, G., Kim, M.G., Park, J.Y., Oh, Y.K.: Graphene-based nanosheets for delivery of chemotherapeutics and biological drugs. Adv. Drug Deliv. Rev. **105**, 205–227 (2016). https://doi.org/10.1016/j.addr.2016.04.004
43. Yao, J., Sun, Y., Yang, M., Duan, Y.: Chemistry, physics and biology of graphene-based nanomaterials: new horizons for sensing, imaging and medicine. J. Mater. Chem. **22**, 14313 (2012). https://doi.org/10.1039/c2jm31632c
44. McCallion, C., Burthem, J., Rees-Unwin, K., et al.: Graphene in therapeutics delivery: problems, solutions and future opportunities. Eur. J. Pharm. Biopharm. **104**, 235–250 (2016). https://doi.org/10.1016/j.ejpb.2016.04.015
45. Barahuie, F., Saifullah, B., Dorniani, D., et al.: Graphene oxide as a nanocarrier for controlled release and targeted delivery of an anticancer active agent, chlorogenic acid. Mater. Sci. Eng. C **74**, 177–185 (2017). https://doi.org/10.1016/j.msec.2016.11.114
46. Yang, Y., Asiri, A.M., Tang, Z., et al.: Graphene based materials for biomedical applications. Mater. Today **16**, 365–373 (2013). https://doi.org/10.1016/j.mattod.2013.09.004
47. Zhang, P., Qiao, Z.-A., Dai, S.: Recent advances in carbon nanospheres: synthetic routes and applications. Chem. Commun. **51**, 9246–9256 (2015). https://doi.org/10.1039/C5CC01759A
48. Nieto-Marquez, A., Romero, R., Romero, A., Valverde, J.L.: Carbon nanospheres: synthesis, physicochemical properties and applications. J. Mater. Chem. **21**, 1664–1672 (2011). https://doi.org/10.1039/C0JM01350A
49. Mochalin, V.N., Shenderova, O., Ho, D., Gogotsi, Y.: The properties and applications of nanodiamonds. Nat. Nanotechnol. **7**, 11–23 (2011). https://doi.org/10.1038/nnano.2011.209
50. Huang, H.-C., Barua, S., Sharma, G., et al.: Inorganic nanoparticles for cancer imaging and therapy. J. Control Release **155**, 344–357 (2011). https://doi.org/10.1016/j.jconrel.2011.06.004
51. Brouns, S.J.J., Jore, M.M., Lundgren, M., et al.: Linked references are available on JSTOR for this article: small CRISPR RNAs guide antiviral defense in prokaryotes. Science **321**, 960–964 (2016)
52. Luk, B.T., Zhang, L.: Cell membrane-camouflaged nanoparticles for drug delivery. J. Control Release **220**, 600–607 (2015). https://doi.org/10.1016/j.jconrel.2015.07.019
53. Millan, C.G., Marinero, M.L.S., Castaneda, A.Z., Lanao, J.M.: Drug, enzyme and peptide delivery using erythrocytes as drug carrier. J. Control Release **95**, 27–49 (2004). https://doi.org/10.1016/j.jconrel.2003.11.018
54. Hu, C.-M.J., Zhang, L., Aryal, S., et al.: Erythrocyte membrane-camouflaged polymeric nanoparticles as a biomimetic delivery platform. Proc. Natl. Acad. Sci. **108**, 10980–10985 (2011). https://doi.org/10.1073/pnas.1106634108
55. Dehaini, D., Fang, R.H., Zhang, L.: Biomimetic strategies for targeted nanoparticle delivery. Bioeng. Transl. Med. **1**, 30–46 (2016). https://doi.org/10.1002/btm2.10004
56. Tan, S., Wu, T., Zhang, D., Zhang, Z.: Cell or cell membrane-based drug delivery systems. Theranostics **5**, 863–881 (2015). https://doi.org/10.7150/thno.11852
57. Parodi, A., Quattrocchi, N., van de Ven, A.L., et al.: Synthetic nanoparticles functionalized with biomimetic leukocyte membranes possess cell-like functions. Nat. Nanotechnol. **8**, 61–68 (2012). https://doi.org/10.1038/nnano.2012.212
58. Gay, L.J., Felding-Habermann, B.: Contribution of platelets to tumour metastasis. Nat. Rev. Cancer **11**, 123–134 (2011). https://doi.org/10.1038/nrc3004

59. Gao, C., Lin, Z., Jurado-Sánchez, B., et al.: Stem cell membrane-coated nanogels for highly efficient in vivo tumor targeted drug delivery. Small 4056–4062 (2016). https://doi.org/10.1002/smll.201600624
60. Sun, H., Su, J., Meng, Q., et al.: Cancer cell membrane-coated gold nanocages with hyperthermia-triggered drug release and homotypic target inhibit growth and metastasis of breast cancer. Adv. Funct. Mater. **27** (2017). https://doi.org/10.1002/adfm.201604300
61. Gujrati, V., Kim, S., Kim, S.H., et al.: Bioengineered bacterial outer membrane vesicles as cell-specific drug-delivery vehicles for cancer therapy. ACS Nano **8**, 1525–1537 (2014). https://doi.org/10.1021/nn405724x
62. Rynda-Apple, A., Dobrinen, E., McAlpine, M., et al.: Virus-like particle-induced protection against MRSA pneumonia is dependent on IL-13 and enhancement of phagocyte function. Am. J. Pathol. **181**, 196–210 (2012). https://doi.org/10.1016/j.ajpath.2012.03.018
63. Li, Y., Wang, F., Cui, H.: Peptide-based supramolecular hydrogels for delivery of biologics. Bioeng. Transl. Med. **1**, 306–322 (2016). https://doi.org/10.1002/btm2.10041
64. Pacheco, S., Kanou, T., Fung, S.Y., et al.: Formulation of hydrophobic therapeutics with self-assembling peptide and amino acid: a new platform for intravenous drug delivery. J. Control Release **239**, 211–222 (2016). https://doi.org/10.1016/j.jconrel.2016.08.038
65. Koutsopoulos, S., Unsworth, L.D., Nagai, Y., Zhang, S.: Controlled release of functional proteins through designer self-assembling peptide nanofiber hydrogel scaffold. Proc. Natl. Acad. Sci. **106**, 4623–4628 (2009). https://doi.org/10.1073/pnas.0807506106
66. Hauser, C.A.E., Zhang, S.: Designer self-assembling peptide nanofiber biological materials. Chem. Soc. Rev. **39**, 2780 (2010). https://doi.org/10.1039/b921448h

Targeted Phytochemical-Conjugated Gold Nanoparticles in Cancer Treatment

Menka Khoobchandani, Kavita K. Katti, Alice Raphael Karikachery, Velaphi C. Thipe, Pierce L. R. Bloebaum and Kattesh V. Katti

Abstract Cancer is one of the most aggressive diseases whose prognosis remains bleak with current therapies. The tumor microenvironment plays a significant role in the proliferation and invasion of tumor cells. In the complex tumor microenvironment, there are several types of immune-suppressing infiltrating cells, which surround the tumor and promote tumor growth and metastasis. Prostate and pancreatic cancers display tumor-infiltrating immune cells, which differentiate into cells that promote each step of the metastatic cascade and therefore are considered as novel targets for therapy. Even though several chemotherapeutic drugs are in current use for treating these cancers, tumor microenvironment targeting immunotherapeutic drugs, with minimum systemic toxic effects, are needed to save human lives from various forms of cancers. We hypothesize a safe and effective nanoinitiated

M. Khoobchandani (✉) · K. K. Katti · A. R. Karikachery
Velaphi C. Thipe · K. V. Katti
Department of Radiology, University of Missouri,
One Hospital Drive, Columbia, MO 65212, USA
e-mail: khoobchandanim@health.missouri.edu

Pierce L. R. Bloebaum · K. V. Katti
Departments of Physics, University of Missouri,
223 Physics Building, Columbia, MO 65211, USA

Pierce L. R. Bloebaum
Department of Mathematics, University of Missouri,
202 Math Sciences Building, 810 East Rollins Street, Columbia, MO 65211, USA

Velaphi C. Thipe
Department of Chemistry, University of Missouri,
125 Chemistry Building, Columbia, MO 65211, USA

M. Khoobchandani · K. K. Katti · A. R. Karikachery
Velaphi C. Thipe · Pierce L. R. Bloebaum
K. V. Katti
Institute of Green Nanotechnology, University of Missouri,
One Hospital Drive, Columbia, MO 65212, USA

K. V. Katti
University of Missouri Research Reactor (MURR), University of Missouri,
1513 Research Park Drive, Columbia, MO 65211, USA

M. Khoobchandani and A. Saxena (eds.), *Biotechnology Products in Everyday Life*,
EcoProduction, https://doi.org/10.1007/978-3-319-92399-4_3

drug delivery system—'Phytochemical-conjugated gold nanoparticles,' which can inhibit the growth and proliferation of tumor cells without affecting normal cells and at the same time boost the capability of immune system to fight cancer. There are unprecedented opportunities to advance the treatment of cancer using nanotechnology. Among the vast majority of nanoparticles, gold nanoparticles show unique properties, such as large surface-to-volume ratio, small size in the nanorealm, tunable surface chemistry, and capable of drug encapsulation. The controlled and targeted drug delivery approach makes nanoparticles very selective and effective toward cancer growth inhibition. In this chapter, we discuss the synthesis and characterization of clove phytochemical-conjugated gold nanoparticles (CP-AuNPs) and their anticancer activity against prostate (PC-3) and pancreatic cancer (PANC-1) cell lines. The synthesis and full characterization of AuNPs were confirmed by various techniques including UV–Visible spectrophotometry, transmission electron microscopy (TEM), and dynamic light-scattering (DLS) technique. We confirmed that AuNP treatment resulted in decreased cancer cell viability and enhancement in cellular internalization. These nanoparticles will navigate better to the complex tumor sites. This 'Green Nanotechnology' approach will be potentially effective compared to the current conventional cancer therapies.

Keywords Green Nanotechnology · Phytochemicals · Gold nanoparticles Cellular internalization · PC-3 cells · PANC-1 cells

1 Introduction

Cancer, due to its complexity, leads to higher morbidity and mortality in the world. More than 10 million new cases are reported each year [1]. Prostate and pancreatic cancers display the most aggressive solid tumors, as current therapies are not very effective [2]. According to the American Cancer Society, deaths arising from cancer constitute 2–3% of the annual deaths recorded worldwide [2, 3]. Several chemotherapeutic agents are used to treat cancer, but often cause severe toxicity, which dissuades their usage worldwide [4].

However, commonly used treatment methods including chemotherapy, radiation, and surgery have severe drawbacks and are not cancer cell-specific [4, 5]. Most chemotherapeutic drugs kill not only cancer cells but also damage normal cells leading to long-term side effects [4, 6]. Despite significant progress made in developing more sophisticated cancer therapeutic modalities, multidrug resistance continues to be a major complication in chemotherapy [7]. Moreover, cancer treatment has become very expensive and unaffordable to the general population across the world [8]. Therefore, there is an urgent unmet clinical need to develop the next generation of drugs that are target-specific, delivered efficiently with minimal side effects and without rendering the disease drug-resistant [9]. In this context, alternative and/or complementary methods of treatment are poised to be the treatment choices of the future for many cancer patients [9].

In this chapter, we focus on the production of a personalized drug through Green Nanotechnology approach—an innovative strategy pioneered by Katti et al. which makes use of the strong antioxidant property of naturally occurring phytochemicals for both nanoparticle production and to render a myriad of phytochemical-encapsulated nanoparticles with tumor-specific cancer killing characteristics [10]. Nanotechnology is defined as the ability to fabricate, characterize, and manipulate artificial structures at nanoscales. Nanotechnology has the power to radically change methods of cancer treatment [10]. Currently, there are enormous research efforts underway for the design and development of novel nanodevices capable of detecting cancer at an early stage, and also for delivering anticancer drugs specifically to malignant cells. The unique size-dependent characteristics of nanoparticle-based drugs (referred to as 'nanoceuticals') turn nanomaterials to be both effective and selective [11].

Typically, nanoparticles are defined as microscopic particles between 1–100 nm and in some instances particle sizes of up to 1 micron are also considered [11]. Nanoparticulate delivery systems provide better penetration of therapeutic and diagnostic cargo at a reduced risk in comparison with conventional cancer therapies [12]. The nanoparticle distribution within the body is based on various parameters such as their size resulting in longer circulation times and their ability to internalize through tumor cell receptors [13].

We discuss an innovative Green Nanotechnology approach utilizing clove phytochemicals for the development of biocompatible tumor-specific gold nanoparticles. Clove, a flower bud, belongs to the family Myrtaceae '*Syzygium aromaticum.*' It is a very popular spice used in Asian countries. Clove is native to Indonesia and is also grown in several parts of the world. Clove is a common ingredient in Asian, African, and Middle Eastern cuisine. It is widely known for its various pharmacological (antioxidant, anticancer, antimicrobial) properties [14]. Clove is a rich source of polyphenolic compounds such as eugenol, eugenol acetate, and gallic acid—all with highly efficient anticancer properties [14].

We report herein: (i) the production and full characterization of clove phytochemical-functionalized gold nanoparticles (CP-AuNPs) and eugenol-conjugated gold nanoparticles (E-AuNPs); (ii) in vitro stability studies (iii) detailed in vitro cytotoxicity and tumor cellular uptake studies. This work demonstrates the utility of plant-based phytochemicals, through Green Nanotechnology, toward the development of tumor-specific cancer therapeutic agents.

2 Methodology

Materials Eugenol and MTT (3-(4,5-Dimethylthiazol-2-yl)-2,5-Diphenyltetrazolium Bromide) dye were obtained from Sigma (St. Louis, MO, USA). RPMI 1640, DMEM, fetal calf serum and TryplE, Trypan blue, and DAPI (4′,6-diamidino-2-phenylindole) were obtained from Thermo Fisher Scientific,

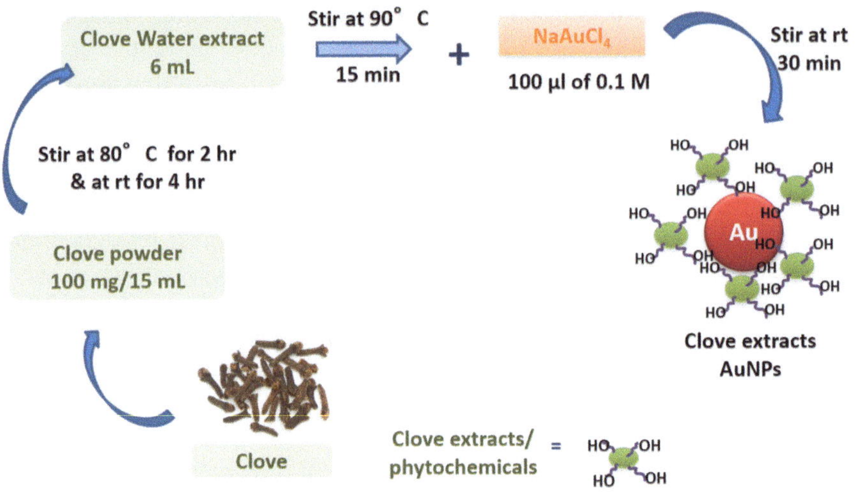

Fig. 1 Synthesis of clove phytochemical-based gold nanoparticles

USA. $NaAuCl_4{\cdot}2H_2O$ (sodium tetrachloroaurate (III) dihydrate) was procured from Alfa Aesar, USA. Double-distilled water (DI) was used throughout the experiment.

Preparation of clove powder Clove was purchased from a local grocery shop. Clove powder was prepared by grinding clove in a mixer. 15 ml of DI water was added to 100 mg clove powder in a scintillation vial. The aqueous solution of clove was sonicated for 15 min and was stirred for 2 h at 80 °C further stirred at room temperature (rt) for another 4 h to obtain the essential phytochemicals in the water. The extract was cooled down to room temperature and filtered off through a Whatman filter paper. It was then stored at 4 °C and used for subsequent gold nanoparticle synthesis.

Synthesis of clove phytochemical-coated gold nanoparticles (CP-AuNPs) The clove extract obtained from the above-mentioned procedure, 6 ml of extract (containing 40 mg of clove phytoextract), was added in a glass vial and stirred for 15 min at 90 °C. Then, 100 μl of 0.1 M $NaAuCl_4$ solution was added to the reaction mixture. The color of the reaction mixture turned to ruby red within few seconds, indicating the formation of gold nanoparticles (Fig. 1). The gold nanoparticles were characterized by UV-visible absorption spectrophotometry, DLS, and TEM analysis. Nanoparticles were stored at 4 °C for further biological studies.

Utility of the active phytochemical 'eugenol' in clove for the synthesis of gold nanoparticles (E-AuNPs). Gold nanoparticles were prepared by mixing eugenol in acetone–water solution in a mild alkaline condition [15]. Briefly, 500 μl of eugenol was dissolved in 20 ml of acetone followed by the addition of 10 ml of water and 100 μl of 0.1 M sodium hydroxide (Solution A). Separately, 10 ml of 2.5 mM aqueous solution of $NaAuCl_4$ was prepared followed by the addition of 100 μl of 0.1 M sodium

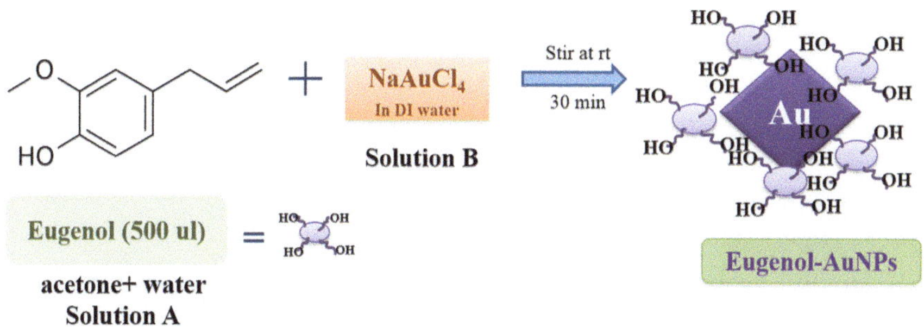

Fig. 2 Synthesis of eugenol (active phytochemical of clove)-based gold nanoparticles (E-AuNPs)

hydroxide (Solution B). Both the solutions were cooled down to 4 °C. Then, solution A was stirred on a magnetic stirrer followed by addition of solution B at room temperature for 30 min. The color of the reaction was changed to a purple color solution, indicating the formation of gold nanoparticles (Fig. 2). After 30 min, 200 μl of 0.1 M sodium hydroxide was added to stabilize the formed polymer eugenol gold colloid. The nanoparticles were stirred for another 2 h to form uniform gold nanoparticles. Nanoparticles were characterized by UV-visible absorption spectrophotometry, particle size analyzer, and TEM analysis. Nanoparticles were stored at 4 °C for further use.

Cell line The human prostate (PC-3) and pancreatic (PANC-1) cancer cell lines were obtained from the American Type Culture Collection (ATCC; Manassas, VA) and cultured by the University of Missouri Cell and Immunobiology Core facility using procedures recommended by ATCC.

Characterization of nanoparticles Transmission electron microscopy (TEM) images were obtained on a JEOL 1400 TEM (JEOL, LTE, Tokyo, Japan). The absorption measurements were obtained by UV–Visible spectrophotometer (Varian Cary 50 conc, USA). The hydrodynamic size and zeta potential were measured using Zetasizer Nano S90 (Malvern Instruments Ltd., USA).

In vitro stability study The in vitro stability of were confirmed by mixing of 1 ml CP-AuNPs to 0.5 ml aqueous solutions of 1% NaCl, 0.5% cysteine, 0.2 M histidine, 0.5% human serum albumin (HSA), 0.5% bovine serum albumin (BSA), and pH7 buffer solution separately. The stability of the conjugates was measured by monitoring the UV–Visible absorbance over a period of 1, 4, 24, 48 h and 1 week. A negligible change in UV–Visible plasmon band confirmed the retention of nanoparticulate composition in all mixtures.

Dark-field microscopy The in vitro cellular internalization analysis of AuNPs was performed by dark-field microscopy techniques. Ultra clean and sterile cover slips were kept in six-well plates. The PC-3 and PANC-1 (8×10^5 cells/ml medium) were seeded into six-well plates in RPMI/DMEM medium separately and incubated for

24 h in CO_2 incubator at 37 °C. CP-AuNPs (100 μl/ml) were added to cells followed by 4 h of incubation at 37 °C. The cells were washed 10–12 times with 1xPBS and fixed with 4% paraformaldehyde (PFA). Cells were further washed 2–3 times with 1xPBS. Slides were prepared by using DAPI nuclear dye and observed with CytoViva dark-field microscope coupled with dual-mode fluorescence. Cell morphology was initially observed, followed by uptake of nanoparticles. Images were captured via Dage Imaging Software.

Cell viability assay The in vitro cytotoxicity of CP-AuNPs was performed as described by the supplier Sigma, USA. Briefly, human cancer (PC-3 and PANC-1) cell line (4×10^4 cells/ml) was plated in 96-well tissue culture plates. The cells were allowed to grow in a CO_2 incubator (37 °C, 5% CO_2, 90% RH) for 24 h. The medium was replaced with the medium containing different dilution of clove extract and CP-AuNPs (80–330 ppm) separately again for 48 and 72 h in CO_2 incubator at 37 °C. MTT dye solution (10 μl/well) was added in the cell suspension and incubated for 4 h in the CO_2 incubator. Periodically viewing the cells under an inverted microscope for the presence of intracellular punctuates showed purple coloration. The reaction was stopped by adding solubilizing DMSO buffer (100 μl) to each well. The plates were kept for 30 min in dark at 25 °C to dissolve all the crystals, and the intensity of the developed color was measured using a microplate reader (SpectraMax plate reader, USA) operating at 570 nm wavelength. Percent of cell viability was calculated by using the formula: $(T/C) \times 100$, where C = absorbance of the control, T = absorbance of treatment. The IC50 values of test samples were also calculated.

Statistical analysis All experimental data are given as mean ± SEM. Statistical analysis was carried out using the one-way analysis of variances (ANOVA) using GraphPad Prism software. $P < 0.05$ was considered significant.

3 Results and Discussion

The aqueous clove extract, rich in antioxidant phytochemicals such as eugenol, was used as a reducing agent to produce the gold nanoparticles. The nanoparticles were synthesized by using Green Nanotechnology process [10]. The nanoparticles were characterized by UV-visible spectrophotometry, DLS, and TEM analysis. Figure 3 represents the UV-visible spectra of the CP-AuNPs that confirmed the surface plasmon resonance (SPR) at 530 ± 2 nm. The SPR indicated that the nanoparticles are smaller in size and uniform based on its narrow peak. The core size of AuNPs (20 ± 5 nm) was measured by TEM (Fig. 3). The hydrodynamic size of the AuNPs was confirmed by DLS measurement to be 70 ± 5 nm (Table 1). These results indicate that AuNPs have a 50 ± 5 nm coating of clove extract around an average core size of 20 ± 5 nm. The stability of the AuNPs was confirmed by measuring its zeta potential, which was −28 ± 3 mV. A high negative charge indicates that the AuNPs are stable and have a long shelf life. The magnitude of zeta potential also provides an

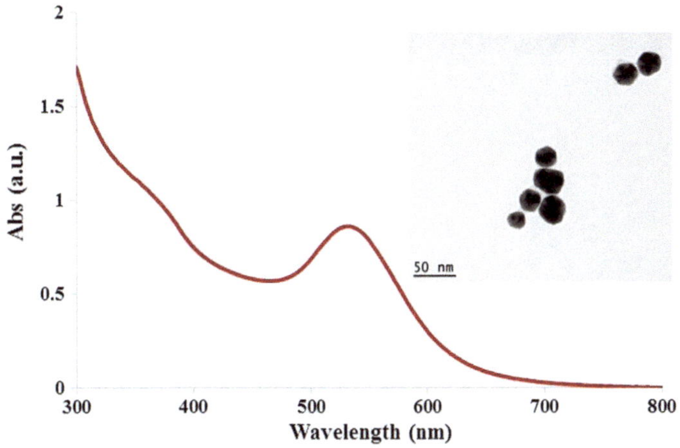

Fig. 3 UV-visible spectrum of CP-AuNPs. Inset TEM image of CP-AuNPs

Table 1 Physicochemical data parameters of gold nanoparticles (AuNPs)

Sample	Absorbance (nm)	Core size by TEM (nm)	Hydrodynamic size by DLS (nm)	Zeta potential (mV)
CP -AuNPs	530 ± 2	20 ± 5	70 ± 5	−28 ± 3
E-AuNPs	540, 706	a = ~35 ± 5 b = ~70 ± 5	60 ± 30	−40 ± 5

Note The core size calculation for E-AuNPs has been discussed below

excellent index for knowing the efficient dispersion and the long-term stability (both in vitro and in vivo) of the CP-AuNPs for biomedical and therapeutic applications.

Eugenol is an active phytochemical in clove extract [14]. The E-AuNPs were characterized by a combination of techniques including UV-visible spectrophotometry, DLS, and TEM. The UV-visible spectrophotometric analysis confirmed the SPR of AuNP at 540 and 706 nm due to diamond- and diamond-chain-shaped nanoparticles corresponding to the transverse and the longitudinal resonance modes (Fig. 4). The core size of E-AuNPs (a = ~35 ± 5 and b = ~70 ± 5 nm) was measured by TEM (Table 1, Fig. 4). These results suggested the successful synthesis of E-AuNPs. The results obtained by DLS measurement revealed that E-AuNPs showed a hydrodynamic size of 60 ± 30 nm and a zeta potential of −40 ± 5 mV.

Calculation of average number of gold atoms per nanoparticle (CP-AuNPs) The average number of gold atoms per nanoparticle is calculated using core size value of the CP-AuNPs obtained from TEM analysis. The average core diameter of the particles (D, nm) is summarized in Table 1. Assuming a spherical shape and a uniform fcc (face-centered cubic) structure, the average number of gold atoms (N) for each nanosphere was calculated using Eq. 1, where ρ is the density of the *fcc* gold (19.3 g/cm^3), M is the atomic weight of gold (197 g/mol), and N_A is the Avogadro number [16].

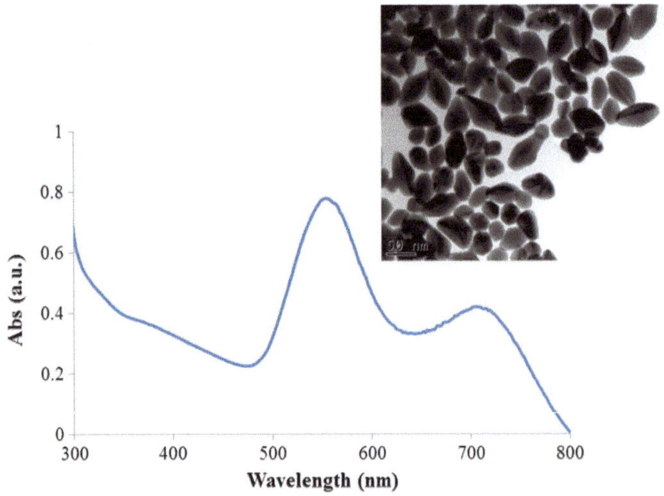

Fig. 4 UV-visible spectrum of E-AuNPs. Inset TEM image of E-AuNPs

$$N = \frac{\pi \rho D^3}{6M} N_A \tag{1}$$

$N = 247{,}200$ gold atoms per nanoparticles.

The calculations above confirmed that CP-AuNPs consisted of ~247,200 gold atoms per nanoparticle.

Determination of molar concentration of nanoparticle solution (CP-AuNPs) The molar concentration of the nanosphere solution was calculated based on the amount of $NaAuCl_4$ used for the nanoparticle synthesis. By dividing the total number of gold atoms (N_T) which is equivalent to the amount of gold utilized (327 ppm, 16.6 μM) over the average number of gold atoms per nanosphere ($N = 247{,}200$) according to Eq. (1), where V (1 lit) is the volume of the reaction solution in liter and N_A is the Avogadro number [16], the concentration of the nanoparticles can be calculated with the following formula:

$$\begin{aligned} C &= \frac{N_T}{N V N_A} \\ &= 6.7 \times 10^{-9} \text{ number of } NPs/\text{mL} \end{aligned} \tag{2}$$

Calculation of average number of gold atoms per nanoparticle (E-AuNPs) The average number of gold atoms per nanoparticle is calculated using core size value of eugenol AuNPs from TEM analysis (Fig. 4, Table 1). The TEM images of the AuNPs synthesized are shown to have an isosceles rhomboid shape in the 2D plane, where a and b describe the length (nm) and width (nm) of the diagonals, respectively.

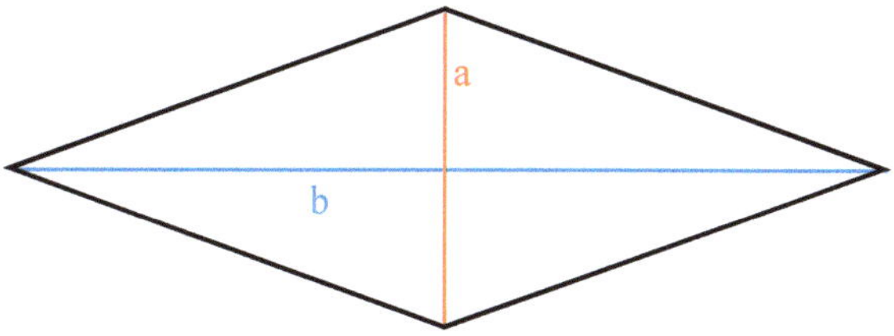

If we consider the 3D shape to be an elongated octahedron, then we can break this into two congruent square pyramids. Assuming the eugenol AuNPs are this shape and the Au atoms maintain a uniform *fcc* lattice structure, the average number of Au atoms/NP (N) can be calculated.

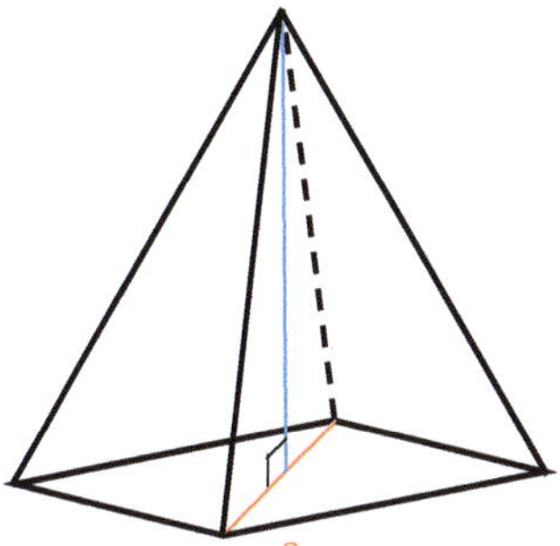

Based on the geometry of a square, the side length of the base $x = a/\sqrt{2}$ and the height $h = b/2$. The volume of these two pyramids V is given below:

$$V = 2V_{\text{pyramid}} = \frac{2}{3}x^2 h = \frac{a^2 b}{6} \tag{3}$$

Since the number of gold atoms per unit cell, $n = 4$, we can calculate N by the equation shown below:

$$N = n\frac{V}{V_c} = 4 * \frac{\frac{a^2 b}{6}}{0.06787\,\text{nm}^3} = 9.823 a^2 b\,\text{nm}^{-3} \tag{4}$$

where V_C is the volume of the *fcc* gold unit cell. From TEM images, the eugenol AuNP core has a width, a ~35 ± 5 nm, and a length, b ~70 ± 5 nm. Using these values in Eq. 4, we obtain the following:

$$N = 9.823 * (35)^2 * 70 = 842,297\,\text{Au atoms} \tag{5}$$

Therefore, we can approximate the number of gold atoms in the eugenol AuNP to be ~842,297 atoms.

Calculation of number of eugenol molecules per gold nanoparticle. The average number of eugenol molecules per gold nanoparticle (E-AuNPs) is calculated using the number of gold nanoparticles per ml (C*) and the number of eugenol molecules per ml (*E*). To find the number of gold nanoparticles per ml, a modified version of the formula to calculate the molar concentration is used, as shown below [16]:

$$C^* = \frac{N_T}{NV} = \frac{1.51 \times 10^{19} \text{Au atoms}}{842,297 \text{Au atoms}/NP * 40.9 \text{ml}} = 4.4 \times 10^{17} \text{NPs/ml} \quad (6)$$

where N_T is the total number of gold atoms initially added to the reaction, which can be calculated from the 10 ml of 2.5 mM gold salt, and V is the total volume of the reaction. To find E (eugenol), we assume that all the eugenol molecules used in the reaction conjugate equally onto the gold nanoparticles. Using that there is 500 μl of eugenol added initially, it is easy to calculate E. As such:

$$E \text{ per AuNP} = \frac{E}{C^*} = \frac{1.944 \times 10^{21} \text{ Eugenol molecules/ml}}{4.37 \times 10^{17} \text{ NPs/ml}} = 4,447,858 \quad (7)$$

Thus, there are approximately 4,447,000 eugenol molecules per gold nanoparticle.

In vitro stability study of CP-AuNPs The stability of pharmaceutical formulation is a critical factor in utilizing it for both in vitro and in vivo biological applications. The stability of the AuNPs was tested by incubating with various biological solutions such as 0.5% cysteine, 0.2 M histidine, 0.5% human serum albumin (HSA), 0.5% bovine serum albumin (BSA), 1% NaCl, and pH 7 buffer solution so as to mimic the physiological conditions. After incubation, the SPR was recorded at various time points (Fig. 5 represents data at one-week time point). The AuNPs SPR peak was unchanged at 530 nm after challenging with various biological fluids, indicating the nanoparticle stability for over a week within all biological solutions. These results indicate that the AuNPs are robust nanoparticles with optimum stability for in vitro and in vivo biological applications.

Cellular uptake The interaction between cancer cells (PC-3 and PANC-1) and CP-AuNPs was demonstrated by dark-field microscopy for exploring the mode of endocytosis. To elucidate the uptake of AuNPs, cells were incubated with different dilutions of AuNPs for 4 h. The results showed that AuNPs internalized inside the cells within 4 h (Fig. 6). The CytoViva technique provides visual observation of the AuNPs internalization in cancer cells and also confirmed significant uptake of AuNPs.

Antitumor efficacy studies through in vitro MTT assay The cell viability profile of CP-AuNPs and clove phytochemicals was evaluated against PC-3 and PANC-1 cells by MTT assay. The cells were treated with different concentrations of test samples. Measurements of cell viability are presented in Figs. 7 and 8. Cell viability profiles demonstrated that AuNPs exhibited dose-dependent efficacy in both

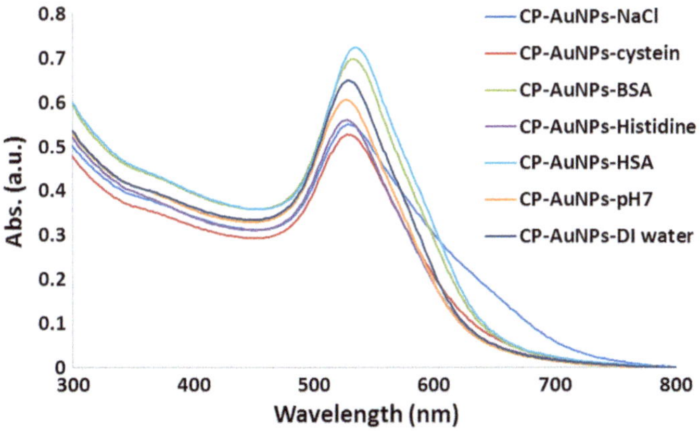

Fig. 5 UV–Visible spectra showing the in vitro stability of CP-AuNPs in aqueous solutions after 1-week incubation

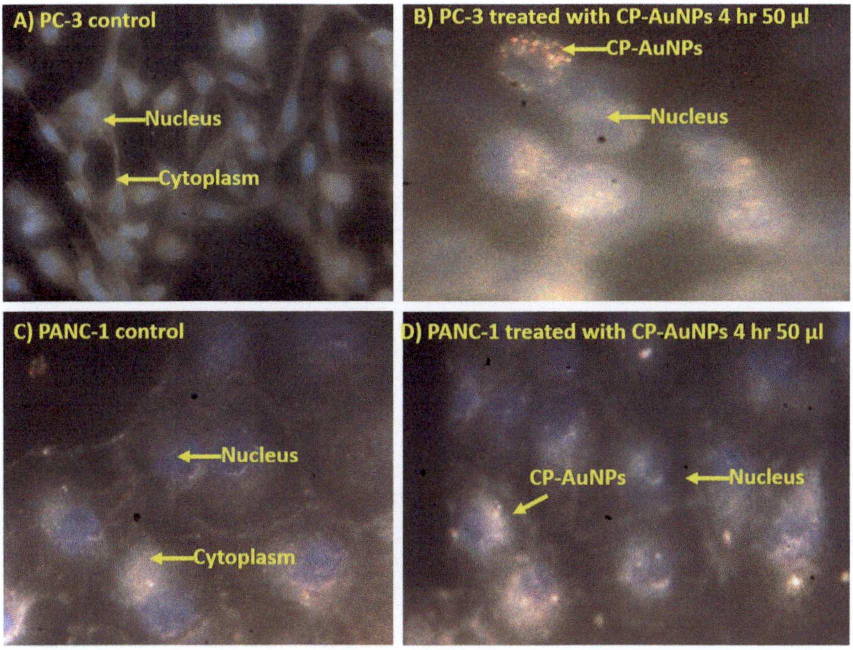

Fig. 6 Dark-field images showing nanoparticle uptake into PC-3 and PANC-1 cells, 4-h post-treatment. (**a**, **b**) PC-3 cells. (**c**, **d**) PANC-1 cells; treated with 100 µl of AuNPs/ ml; and control no treatment

types of tumor cells. This experiment suggests that encapsulating of clove phytochemicals onto gold nanoparticles certainly enhances its pharmacological potential and increases its bioavailability. Figure 7 shows that clove phytochemical-coated

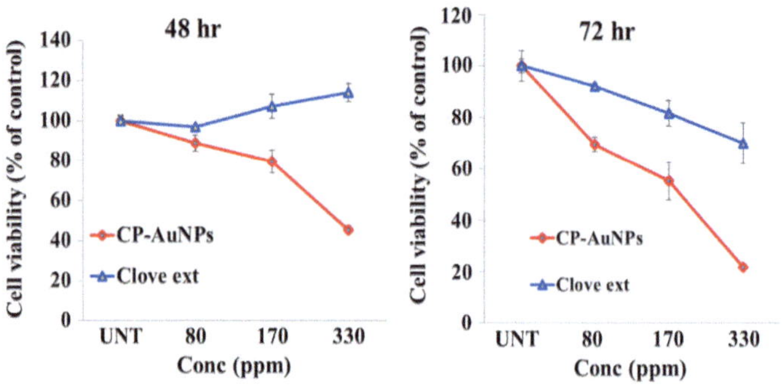

Fig. 7 Effect of CP-AuNPs and clove extract on cell viability of PC-3 cells after 48- and 72-h incubation with increasing concentration. Where UNT is no treatment control group. Concentrations are reported in terms of amount of clove extract in AuNPs and as free phytochemicals

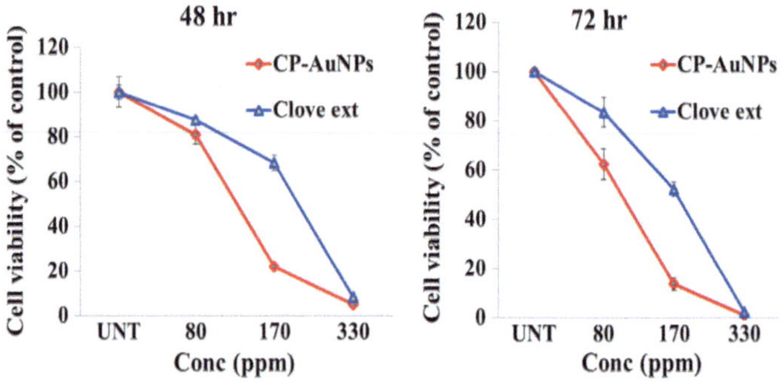

Fig. 8 Effect of CP-AuNPs and clove extract on cell viability of PANC-1 cells after 48- and 72-h incubation with increasing concentration. Where UNT is no treatment control group. Concentrations are reported in terms of amount of clove extract in AuNPs and as free phytochemicals

nanoparticles inhibit 50% cells death at concentration ~300 ppm at 48 h and 170 ppm at 72 h, whereas clove extract showed only 30% cell death at 330 ppm at 72 h against prostate cancer cells.

Figure 8 shows that CP-AuNPs inhibit 50% growth of PANC-1 at concentrations of 100 ppm at 48 h and 80 ppm at 72 h. Under similar experimental conditions, clove extract (without the gold nanoparticles) exhibited 50% inhibition of growth of PANC-1 at concentrations of 300 and 200 ppm at 48 h and at 72 h, respectively. The results corroborated that encapsulation of clove phytochemicals onto AuNPs significantly enhanced their bioefficacy to inhibit PANC-1 cells growth.

These results confirmed that phytochemicals exhibit enhanced antitumor efficacy against various cancer cell lines when encapsulated onto AuNPs using the Green Nanotechnology approach.

Implications of Green Nanotechnology in Oncology Despite numerous investigations demonstrating experimental success in in vitro and preclinical studies, there has been limited progress in the clinical translation of natural phytochemicals and phytochemicals-based vectors to the clinic as the first line or adjuvant therapy in treating cancers and various other diseases. Analysis of extensive in vitro (and in vivo) data provides important following insights on the limitations and impediments on why phytochemicals-based therapy has met with limited success so far: (i) In vitro assays involve direct incubations of cancer cell lines resulting in an acute presentation of phytochemicals. Such exposers induce significant anticancer and antiproliferative effects at tested concentrations which are normally not achieved under normal physiological conditions, of oral delivery, even upon consumption of large quantities of the raw or the pure phytochemicals extracts; (ii) another significant challenge that continues to impede successful clinical translation of various phytochemicals for treating cancers and various diseases/disorders is associated with the problem of bioavailability.

The process of phytochemical disposition in vivo, similar to that of disposition of drugs and other xenobiotics, involves absorption, metabolism, distribution, and excretion, and each of these processes contributes to pharmacokinetic variability and consequently drug efficacies of individual or cocktail of phytochemicals-derived drugs. In the context of designing phytochemicals-based drugs, it is important to recognize that a majority of these phytochemicals are part of normal human diet. This aspect presents a key bioavailability problem that they are efficiently metabolized and cleared by the human body. The suboptimal biological half-life of phytochemicals in vivo means that phytochemical constituents fail to persist in physiological systems, and therefore, therapeutic effects are either usually short-lived or are eliminated. Overall, the enzymatic and non-enzymatic degradation in vivo undermines the beneficial ‘pleiotropic’ effects of phytochemicals in therapy.

Green Nanotechnology approaches of encapsulating hundreds of signatures of clove phytochemical (eugenol) around gold nanoparticles surface, as described in this chapter (Figs. 1 and 2), appear to successfully circumvent problems related to delivery and bioavailability—the two key problems that have haunted the full utility of phytochemicals as first line or adjuvant therapeutic agents. Gold nanoparticles, with multitudes of clove phytochemical (eugenol) molecules bound to the large surface area of gold, deliver a huge cargo of the therapeutic phytochemical through endocytosis as shown in Fig. 6. Entry of clove phytochemical-encapsulated gold nanoparticles is synergistically poised due to the optimum size and charge of gold nanoparticles as well as tumor cell receptor affinity of the bound eugenol—thus leading to selective uptake of therapeutic nanoparticles by tumor cells. The innate antiangiogenesis effects of gold and the inherent therapeutic power of eugenol collectively result in tumor destruction.

Eugenol-functionalized gold nanoparticles in immunotherapy of cancer—concluding remarks The literature indicates that polyphenols present in the clove extract would be effective in the modulation of intracellular nuclear factor-kappaB (NF-κB) signaling pathway involved in the deregulated expression of cell proliferation and cell cycle regulatory molecules [17]. Therefore, our Green Nanotechnology approach which provides a pragmatic pathway for tumor cell-specific delivery of clove phytochemical (eugenol)-encapsulated gold nanoparticles is a scientifically sound approach for tumor therapy through immunomodulatory pathways. The targeting of NF-κB signaling pathway by CP-AuNPs (eugenol) will have a significant impact on the utility of eugenol-functionalized gold nanoparticles as immunomodulatory agents in tumor therapy.

The endocytosis of CP-AuNPs (Fig. 6) would inhibit NF-κB signaling specifically within the tumor-associated macrophages (TAMs) [18]. This transforms TAMs to become cytotoxic to tumor cells by switching to a 'classically' activated M-1 phenotype displaying high levels of tumor suppressive IL-12 and concomitantly reduce tumor proliferating IL-10 cytokine [18]. Therefore, CP-AuNPs (active phytochemical 'eugenol') provide new strategies for immunomodulatory cancer therapy through: (i) specific interference with M2-like TAM survival by selectively inhibiting their signaling cascades, (ii) suppression of macrophage recruitment to tumors, and (iii) re-education of tumor-promoting M2-like TAMs to a tumor suppressing M1-like phenotype.

Acknowledgements We thank the Institute of Green Nanotechnology for supporting this interdisciplinary research. We also thank the University of Missouri Departments of Radiology and School of Medicine for financial support.

Author Contributions Menka Khoobchandani, Kavita K. Katti, and Kattesh V. Katti conceived and designed experiments, analyzed data, and edited the manuscript. Menka Khoobchandani performed the experiments, analyzed and observed the data, and wrote the paper. Kavita K. Katti and Alice Raphael Karikachery edited the manuscript. Velaphi C. Thipe reviewed the manuscript. Pierce L.R. Bloebaum analyzed and performed nanoparticles calculation.

Conflicts of Interest The authors declare no conflict of interest.

References

1. (a) Jemal, A., Ward, E.M., Johnson, C.J., Cronin, K.A., Ma, J., Ryerson, B., Mariotto, A., Lake, A.J., Wilson, R., Sherman, R.L., Anderson, R.N., Henley, S.J. Kohler, B.A., Penberthy, L., Feuer, E.J., Weir, H.K.: J. Natl. Cancer Inst. **109** (2017); (b) Ryerson, A.B., Eheman, C.R., Altekruse, S.F., Ward, J.W., Jemal, A., Sherman, R.L., Henley, S.J., Holtzman, D., Lake, A., Noone, A.M., Anderson, R.N., Ma, J., Ly, K.N., Cronin, K.A., Penberthy, L., Kohler, B.A.: Cancer **122**, 1312–1337 (2016)
2. Heymach, J., Krilov, L., Alberg, A., Baxter, N., Chang, S.M., Corcoran, R., Dale, W., DeMichele, A., Magid Diefenbach, C.S., Dreicer, R., Epstein, A.S., Gillison, M.L., Graham, D.L., Jones, J., Ko, A.H., Lopez, A.M., Maki, R.G., Rodriguez-Galindo, C., Schilsky, R.L., Sznol, M., Westin, S.N., Burstein, H.: J. Clin. Oncol. **36**, 1020–1044 (2018)
3. Moeller, B., Balagamwala, E.H., Chen, A., Creach, K.M., Giaccone, G., Koshy, M., Zaky, S., Rodrigues, G.: Pract. Radiat. Oncol. (2018)

4. (a) Kobayashi, S., Ueno, M., Hara, H., Irie, K., Goda, Y., Moriya, S., Tezuka, S., Tanaka, M., Okusaka, T., Ohkawa, S., Morimoto, M.: Oncology **91**, 117–126 (2016); (b) Gardiner, R.F., Nicol, D.L., Green, H.J., Pakenham, K.I.: BJU Int. **98**, 229–230 (2006); (c) McCann, J.: J. Natl. Cancer Inst. **92**, 449–451 (2000)
5. Arriagada, R., Averbeck, D., Dahl, A.A., Darby, S., Fossa, S., Friberg, S., Polgar, C., de Vathaire, F.: Eur. J. Cancer **45**, 354–359 (2009)
6. Slonina, D., Klimek, M., Szpytma, T., Gasinska, A.: Int. J. Radiat. Biol. **76**, 1255–1264 (2000)
7. Chang, H.P., Lu, C.C., Chiang, J.H., Tsai, F.J., Juan, Y.N., Tsao, J.W., Chiu, H.Y., Yang, J.S.: Int. J. Oncol. (2018)
8. (a) Kaye, D.R., Min, H.S., Herrel, L.A., Dupree, J.M., Ellimoottil, C., Miller, D.C.: Oncologist (2018); (b) Bao, Y., Maciejewski, R.C., Garrido, M.M., Shah, M.A., Maciejewski, P.K., Prigerson, H.G.: J. Pain. Symptom. Manage. **55**, 1113–1121 e1113 (2018)
9. (a) Khoobchandani, M., Katti, K., Maxwell, A., Fay, W.P., Katti, K.V.: Int. J. Mol. Sci. **17**, 316 (2016); (b) Khoobchandani, M., Ganesh, N., Gabbanini, S., Valgimigli, L., Srivastava, M.M.: Fitoterapia **82**, 647–653 (2011); (c) Khoobchandani, M., Ojeswi, B.K., Sharma, B., Srivastava, M.M.: Oxid. Med. Cell Longev. **2**, 160–165 (2009); (d) Ojeswi, B.K., Khoobchandani, M., Hazra, D.K., Srivastava, M.M.: Hum. Exp. Toxicol. **29**, 369–375 (2010); (e) Yue, Q.X., Gao, G.G., Zou, G.Y., Yu, H.Q., Zheng, X.: Biomed. Res. Int. (2017)
10. (a) Al-Yasiri, A.Y., Khoobchandani, M., Cutler, C.S., Watkinson, L., Carmack, T., Smith, C.J., Kuchuk, M., Loyalka, S.K., Lugao, A.B., Katti, K.V.: Dalton Trans. (2017); (b) Chanda, N., Kattumuri, V., Shukla, R., Zambre, A., Katti, K., Upendran, A., Kulkarni, R.R., Kan, P., Fent, G.M., Casteel, S.W., Smith, C.J., Boote, E., Robertson, J.D., Cutler, C., Lever, J.R., Katti, K.V., Kannan, R.: Proc. Natl. Acad. Sci. U S A **107**, 8760–8765 (2010); (c) Chanda, N., Shukla, R., Katti, K.V., Kannan, R.: Nano Lett. **9**, 1798–1805 (2009); (d) Chanda, N., Upendran, A., Boote, E.J., Zambre, A., Axiak, S., Selting, K., Katti, K.V., Leevy, W.M., Afrasiabi, Z., Vimal, J., Singh, J., Lattimer, J.C., Kannan, R.: J. Biomed. Nanotechnol. **10**, 383–392 (2014); (e) Gamal-Eldeen, A.M., Moustafa, D., El-Daly, S.M., Abo-Zeid, M.A.M., Saleh, S., Khoobchandani, M., Katti, K., Shukla, R., Katti, K.V.: Biomed. Pharmacother. **89**, 1045–1054 (2017); (f) Gamal-Eldeen, A.M., Moustafa, D., El-Daly, S.M., El-Hussieny, E.A., Saleh, S., Khoobchandani, M., Bacon, K.L., Gupta, S., Katti, K., Shukla, R., Katti, K.V.: J. Photochem. Photobiol. B **163**, 47–56 (2016); (g) Katti, K., Chanda, N., Shukla, R., Zambre, A., Suibramanian, T., Kulkarni, R.R., Kannan, R., Katti, K.V.: Int. J. Green Nanotechnol. Biomed. **1**, B39–B52 (2009); (h) Kattumuri, V., Katti, K., Bhaskaran, S., Boote, E.J., Casteel, S.W., Fent, G.M., Robertson, D.J., Chandrasekhar, M., Kannan, R., Katti, K.V.: Small **3**, 333–341 (2007); (i) Khoobchandani, M., Katti, K., Maxwell, A., Fay, W.P., Katti, K.V.: Int. J. Mol. Sci. 17 (2016); (j) Khoobchandani, A.Z.M., Katti, K., Lin, C.H., Katti, K.V.: Int. J. Green Nanotechnol. **1**, 1–15 (2013); (k) Nune, S.K., Chanda, N., Shukla, R., Katti, K., Kulkarni, R.R., Thilakavathi, S., Mekapothula, S., Kannan, R., Katti, K.V.: J. Mater. Chem. **19**, 2912–2920 (2009); (l) Shukla, R., Chanda, N., Zambre, A., Upendran, A., Katti, K., Kulkarni, R.R., Nune, S.K., Casteel, S.W., Smith, C.J., Vimal, J., Boote, E., Robertson, J.D., Kan, P., Engelbrecht, H., Watkinson, L.D., Carmack, T.L., Lever, J.R., Cutler, C.S., Caldwell, C., Kannan, R., Katti, K.V.: Proc. Natl. Acad. Sci. U S A **109**, 12426–12431 (2012); (m) Shukla, R., Nune, S.K., Chanda, N., Katti, K., Mekapothula, S., Kulkarni, R.R., Welshons, W.V., Kannan, R., Katti, K.V.: Small **4**, 1425–1436 (2008); (n) Silva, R.A., Giordano, R.J., Gutierrez, P.S., Rocha, V.Z., Rudnicki, M., Kee, P., Abdalla, D.S., Puech-Leao, P., Caramelli, B., Arap, W., Pasqualini, R., Meneghetti, J.C., Marques, F.L., Khoobchandani, M., Katti, K.V., Lugao, A.B., Kalil, J.: Int. J. Mol. Sci. 17 (2016)
11. Dreaden, E.C., Austin, L.A., Mackey, M.A., El-Sayed, M.A.: Ther Deliv **3**, 457–478 (2012)
12. (a) Langston Suen, W.L., Chau, Y.: J. Pharm. Pharmacol. **66**, 564–573 (2014); (b) Shan, Y., Ma, S., Nie, L., Shang, X., Hao, X., Tang, Z., Wang, H.: Chem. Commun. (Camb) **47**, 8091–8093 (2011)
13. Nichols, J.W., Bae, Y.H.: Nano Today **7**, 606–618 (2012)
14. Cortes-Rojas, D.F., de Souza, C.R., Oliveira, W.P.: Asian Pac J Trop Biomed **4**, 90–96 (2014)

15. Milczarek, G., Ciszewski, A.: Colloids Surf. B Biointerfaces **90**, 53–57 (2012)
16. Liu, X., Atwater, M., Wang, J., Huo, Q.: Colloids Surf. B Biointerfaces **58**, 3–7 (2007)
17. Tsai, T.H., Huang, W.C., Lien, T.J., Huang, Y.H., Chang, H., Yu, C.H., Tsai, P.J.: Food Agric. Immunol. **28**, 916–931 (2017)
18. Islamuddin, M., Chouhan, G., Want, M.Y., Ozbak, H.A., Hemeg, H.A., Afrin, F.: Plos Neglected Trop. Dis. 10 (2016)

Part II
Intelligent Devices

Enterprising Artificial Intelligence: Aiding Patients as Well as the Healers

Arpita Saxena

Abstract Artificial intelligence is 'the' word that best describes the future of health sector. With the advent of the age of Internet of things (IOT), almost every aspect of life has changed. Biotechnology also does not remain untouched with these changes. Today, while there are robots to generate and analyze the data, there are also devices with ability to interpret those results. As the biotechnologists today extend their learning into making software for medical and clinical data mining, there are also computer and electrical engineers who seek their careers in making smart and intelligent devices which aid to medical, pharmaceutical, or clinical requirements. In the last two decades, especially the last 7–9 years, the market has witnessed many such devices and there is a whole community of start-ups mushroomed in the recent past, dedicated to the making of these devices. This review revisits some of the latest, popular, and enterprising technologies in biotech space, which integrate artificial intelligence with medical science. This chapter also emphasizes on the growing culture of start-ups around the globe dedicated to manufacturing of devices and software that assist the hospital staff with medical data collection, maintenance, and analysis, while most of them foster the field of diagnostics. These several categories of devices have not yet replaced human intelligence completely, though they have speeded up the process and ensured more efficacy level than simple, mechanical, humanized efforts.

Keywords AI · E-Health · M-Health · Machine learning · Start-ups
Deep learning

1 Introduction

As 'Sophia,' the humanoid [1], attains citizenship of a country (now received citizenship of more than one country) and retains many special rights which are not even

A. Saxena (✉)
AAPT Foundation, S-9, Opp. DKMM Homeopathic College, Aurangabad 431004, India
e-mail: arpitaspak@gmail.com

M. Khoobchandani and A. Saxena (eds.), *Biotechnology Products in Everyday Life*,
EcoProduction, https://doi.org/10.1007/978-3-319-92399-4_4

offered to the women citizens of respective country, the debate of artificial verses human intelligence gets heated again [2]. The debating communities on either side of the table, howsoever different principally, appreciate the technology and emotionally expressive face of the humanoid. This picture typically brings to our minds a visualization of human being assisted by artificial intelligence in every sphere of life, to say the least, that is the basic idea behind their creation.

Artificial intelligence (AI) in human health care has been a game changer [3]. The last 15 years were dedicated to smart devices that ensure speed, precision, monitoring, and control over health and diseases. This era has seen a shift in focus from human resource (humanized team efforts) to data resource, from data resource (collection of detailed medical facts and figures) to machine learning (large data pool which starts co-relating these facts and figures), and from machine learning to artificially obtained intelligence (ability to interpret a new sample on the basis of past data analysis) (Fig. 1).

Electronic-health (e-health) solutions are the software applications related to health care that provide tools, processes, and communication means to support electronic healthcare practice [4–7]. Mobile-health (m-health) comes under the umbrella of e-health that refers to e-health functionality on mobile phones, which further diverges into technologies of networking, medical sensors, computation, and communication [8]. There is another term called tele-health which refers to medical counseling on phone by an expert doctor to medical staff in remote areas lacking high-end medical expertise.

Artificial intelligence has recently stepped into health care but has made major transformations to the whole system of diagnosis, analysis, and even drug development.

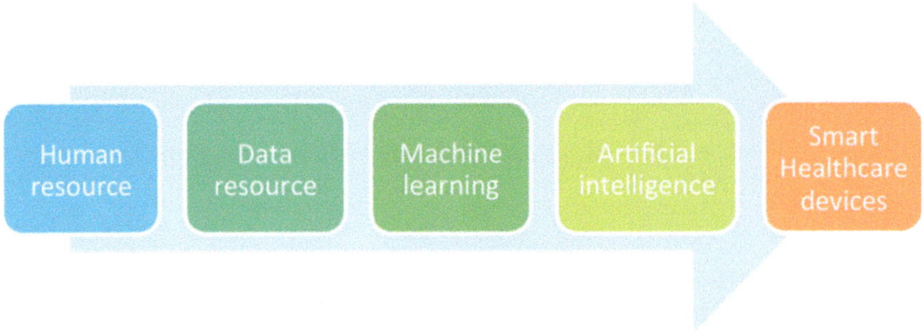

Fig. 1 Evolution of healthcare systems with respect to artificial intelligence

1.1 AI Around the World

The geographical distribution of start-ups in artificial intelligence around the world is itself an interesting topic to work on as the dynamics are changing fast. There are several applications of AI like health care, retail, financial services, manufacturing, [9] and if all included, the USA alone has around 3000 start-ups followed by China, Britain, India, and Canada. Countries like Israel, Germany, France, Spain, and Switzerland also make a mark [10]. Within the USA, states holding maximum start-up sector were California (51%), New York (11%), Massachusetts (9%) and states like Utah, Colorado, New Jersey, Texas, Illinois, Virginia, Washington, Michigan, Ohio, and Florida are witnessing a generation of well-funded start-up growth [11]. Globally, the AI revenue is estimated to reach $37 billion by 2025 [12]. A sole share of $19.2 billon will be claimed by healthcare applications of AI with image analysis leading the chase, followed by virtual assistance, patient data processing, computational drug discovery, and converting paperwork into digital data [13].

1.2 Some Products as an Example of Collaborative Research

Before we get into AI and its applications in medical entrepreneurship, let us witness a few interesting stories where biotechnology has wedded multidisciplinary research and led to wonderful products.

1.3 3D Printed Bones

Rhys Cornock, a 23-year-old graduate from University of Wollongong, Australia, is transforming the world of medical sciences by his 3D printed bones [14]. He has used advanced bioengineering techniques to manufacture complex scaffolds that can exactly fit into fractured bone areas with the help of specific scans and imaging of the fractured area. These perfectly tailored bones, printed through 3D printers, are then transplanted into patients' body.

Another product of the same university is a biopen that is used to deliver live cells and growth factors directly to the site of injury. This reduces the surgery as well as recovery time as regeneration of functional bone and cartilage is stimulated by the growth factors. These types of implants are best for cases with diseased bone structures. It is made from biodegradable, non-toxic material that provides a structure through its polymer component and recovery through its growth factors and drug component [15].

In the same area of research, Professor Milan Brandt, RMIT's lead researcher, furthered the 3D printed bone implants surgery by bringing in the robotics. The project team has combined 3D printing, robotic surgery, and advanced manufacturing to create tailored implants for patients with bone cancer [16]. This is a 5-year project that brings together the Australian Government, RMIT University, the University of Technology Sydney (UTS), St Vincent's Hospital, Melbourne, and global medical technology firm Stryker.

1.4 Sickle Cell Chip

A diagnostic kit developed by a team from IIT Bombay has made it possible to detect sickle cell anemia, a common blood disorder found in tribal areas of central and southern India. This kit is affordable, portable and can be used even by untrained medical staff. Debjani Paul, Ninad Mehendale, and Ammar Jagirdar from the Indian Institute of Technology, Powai, have developed this kit [17] as an example of laboratory on chip diagnostics that can facilitate distant, rural, underprivileged sections of human population. The idea was funded by BIRAC and the Bill & Melinda Gates Foundation. Invention of this kit has proved to be a savior for all the patients, especially children who died before even being diagnosed with the disease. A drop of patient blood added to this—reagents loaded plastic micro-fluidic chip—can be imaged by a mobile phone camera.

Other than these two stories, there are other products like wearable solar cell-based textiles [18, 19] that can be coupled with sensors to form active smart materials that can be used for monitoring health parameters like electrocardiogram (ECG) [20], electromyography (EMG) [21], electroencephalography (EEG) [22, 23]. Another product which is already in the market is a pair of affordable socks from 'Barefoot' that does not need shoes as they are 15 times stronger than steel [24]. It gives a feeling of barefoot walking and gives a fine grip to the feet during any sport.

2 An Overview of Healthcare Products Using AI Around the World

2.1 When Artificial Intelligence Completes Human Body [25]

See Table 1.

2.2 Medical Imaging, Diagnostics, and Artificial Intelligence

See Table 2.

Table 1 List of the various products that have replaced parts of human organs or play as extended body parts, with their advantages over contemporary systems [25]

Products	Features and advantages over existing technologies	Companies/university
(i) Portable dialysis	• A machine that can carry out complete dialysis process efficiently round the clock • Fully automated, battery-operated, waterproof, and absolutely portable • Cuts down long hospital visits	Xcorporeal, Los Angeles [25, 26]
(ii) Absorbable heart stent	• Biodegradable, this stent starts to dissolve after 6 months and completely departs in 18 months, leaving behind a healthy artery • Another category of stents releases medication at prefixed intervals that keeps the artery from narrowing again	Abbott Laboratories, Illinois [25, 27]
(iii) Smart contact lens	• A pair of lens that contain conductive wires that continuously monitor pressure and fluid flow within the eyes of people at risk with glaucoma • The lenses wirelessly transmit information to a computer where doctors can use it • In the future, lenses may also automatically dispense drugs in response to pressure changes	University of California, Davis [25]
(iv) Speech restorer	• This device produces speech through the impulses received wirelessly from its counterpart installed on patient's neck brace • The neuronal signals signaled from the brain to the vocal cords are detected and converted to words though smartphone • Patients have a feel of slowly sounding out words	Illinois-based Ambient Corporation and Texas Instruments [25]

(continued)

Table 1 (continued)

Products	Features and advantages over existing technologies	Companies/university
(v) Autonomous wheelchair	• A mobile and audio-sensitive wheelchair which listens to its user and takes to places • Works on machine learning principles • Builds maps using Wi-Fi • Equipped with cameras, laser rangefinders, and a collision—avoidance system	MIT [25, 28]
(vi) Nerve regenerator	• This is a nanogel that self-assembles into a scaffold of nanofibers • These nanofibers express peptides that instruct stem cells to normally form scar tissue which in turn produce cells that encourage nerve development	Northwestern University [25, 29]
(vii) Stabilizing insoles	• This technology is developed to help elderly patients against falls • These sensors are installed on the insoles to analyze the pressure distribution of the feet	Lieberman's iShoe [25]
(viii) Smart pill	• These pills not only contain medication but are also loaded with sensors that track the exact time drugs are ingested • These signals from sensors are received on Band-Aid-like receivers on the skin • Also used to monitor heart rate and respiration and the data can be wirelessly transmitted to a smart device	Proteus Biomedical, California [30]
(ix) Prosthetic feedback	• This device stretches an amputee's skin near the prosthesis in ways that provide feedback about the limb's position and movement.	Karlin Bark, Stanford University [25]

(continued)

Table 1 (continued)

Products	Features and advantages over existing technologies	Companies/university
(x) Muscle stimulator—MyoSpare	• This device facilitates muscle movement during fracture to keep them strong during recovery • Battery-operated, this device uses electrical stimulators, to be worn underneath casts	StimuHeal, Israel [25]

Table 2 Companies using AI involved in the field of medical imaging and scanning [32, 33, 35]

Products/companies	Features and advantages over existing technologies	Founder
(1) Neuralink 2016	• Developing ultra-high bandwidth brain–machine interfaces to connect humans and computers	Elon Musk based in California, USA
(2) Arterys with GE Healthcare 2015	• Cloud-based cardiac imaging • Fast and accurate, takes 6–10 min instead of an hour • Acquires seven dimensions of data, which include 3D heart anatomy, blood flow rate, and blood flow direction	ViosWorks Project [31]
(3) MJN Neuroservices 2014	• MJN develops high-technology innovative devices to predict epileptic seizures	David Blanquez Caurel based in Catalonia, Spain
(4) 3Scan 2011	• Robotics microscopy used for tissue analysis • Can eliminate some drudgery for drug researchers who have been stuck using manual processes • Faster than any other manual processes	Ben Turley start-up based in Provo, Utah, USA
(5) Imagia's Deep Radiomics 2015	• Uniting radiomics and deep learning • Utilizes existing routine clinical imaging data to discover robust imaging biomarkers that predict cancer patient outcomes	Alexandre Le Bouthillier based in Quebec, Canada

(continued)

Table 2 (continued)

Products/companies	Features and advantages over existing technologies	Founder
(6) Butterfly Network 2011	• MRI and ultrasounds made cheaper convenient and more efficient • Automated medical imaging processes • Alternatively, an iPhone-sized scanner that has to be held up to a person's chest to see a vivid, moving, 3-D image of what's inside is also available • It works with a deep learning algorithm trained by ultrasound experts	Jonathan Rothberg start-up based in Guilford, USA
(7) Bay Labs 2013	• Interprets ultrasounds to treat heart disease. • Software for diagnosis of rheumatic heart disease (RHD) of congenital heart disease	Charles Cadieu Company at San Francisco, CA, USA
(8) Enlitic 2014	• Based on machine learning, its unique interpretation skills are beyond compare • 10,000 times faster than average radiologist with false negative 0%	Felix Baldauf-Lenschen Company at San Francisco
(9) Cancer spit test 2016	• This breakthrough technology detects oral cancer from a single drop of saliva putting away all biopsies and pain • This laboratory on chip technology has sensor which can detect proteins associated with cancer cells • These proteins further react with fluorescent dyes that can be detected with a fluorescent microscope	University of California, Los Angeles

2.3 *AI and Drug Discovery*

See Table 3.

2.4 *AI and Data Mining/Medical Records Mining*

See Table 4.

2.5 *Indian Start-ups in the Healthcare Sector Using AI*

See Table 5.

3 Challenges Involved in Medical Applications of Artificial Intelligence [33]

There are unexpected and unique obstacles in using AI as a medical savior like: (a) Stakeholdership is scattered and so whenever there is an operational error, it is difficult to trace back the responsible factor or person. (b) Sharing health data across hospitals, through mobile devices, or in other databases implies many unique challenges with Health Insurance Portability and Accountability Act (HIPAA, passed by Congress in 1996) compliance. (c) The job of replacing an entire doctor is unlikely as personal diagnosis at the time of consultation cannot be replaced entirely by past data history. (d) Machine learning and deep learning (unlike stodgier AI approaches like expert systems) are unable to express why they achieved the result that they did.

The challenges of this field are ever increasing as they are counter-faced through innovation and further more grit.

4 Discussion

With the societal and lifestyle changes, health monitoring and maintenance itself have become a challenge. The solution also comes from the same age of e-health where artificial intelligence assists human intelligence to confront these challenges of day-to-day life not only for the patients but also for their caretakers. Mobile phones, electronic gadgets, and Internet services have involved non-medical practitioners along with patients themselves, to be the first monitors of their ailments, and have personalized health monitoring and treatment. An ever-increasing number of health-

Table 3 Companies involved in the drug discovery process using AI [32, 33]

Products/companies	Year	Features and advantages over existing technologies	Founders
(1) Atomwise	2012	• First deep learning technology for novel small molecules discovery • Unprecedented speed, accuracy, and diversity • Aims to reduce the costs of medicine development • Total funding amount $6,570,000 • Target disease/organism—Ebola virus and multiple sclerosis	Abraham Heifets, Alexander Levy, Dr. Izhar Wallach
(2) Recursion Pharmaceuticals	2013	• Best elements of high-throughput biology and automation • The company has identified novel uses for known drugs, bioactive compounds, and shelved pharma assets in the space of rare genetic diseases • They promise to fulfill their ambitious goal of curing 100 diseases in just 10 years • Total funding amount $84,330,000	Blake Borgeson, Chris Gibson, Dean Li
(3) Whole Biome	2014	• Focus area—microbiome interventions • For the known fact that microbiome represents a wealth of opportunity in shaping health and house many microbes that are beneficial for health, they have made genomic and analytical tools to make it happen • One of their products help women avoid preterm labor	Colleen and James; collaborator, Mayo Clinic
(4) Deep Genomics	2014	• Focus area—genome • Based on machine learning from data of mutations • Ability to try and predict the effects of a particular mutation • Their findings are used for genome-based therapeutic development, molecular diagnostics, targeting biomarker discovery, and assessing risks for genetic disorders • Total funding amount $16,700,000	Andrew Delong, Brendan Frey, Hannes Bretschneider, Hui Yuan Xiong
(5) iCarbonX	2015	• Maintains a personal biological library containing biological samples such as saliva, proteins, and DNA of individuals • The company is developing algorithms to analyze the data, with the intention of recommending tailored wellness programs, food choices, and possibly prescription medicines • Total funding amount CN¥1,300,000,000	Jun Wang, Yingrui Li

(continued)

Table 3 (continued)

Products/companies	Year	Features and advantages over existing technologies	Founders
(6) Turbine	2015	• Personalized treatments for any cancer type or patient faster than any traditional healthcare service • The technology models cell biology on the molecular level • It can identify the best drug to target a specific tumor with; moreover, it identifies complex biomarkers and design combination therapies by performing millions of simulated experiments each day	Szabolcs Nagy, Daniel Veres, Kristof Szalay, Ivan Fekete

care start-ups are incorporating machine learning and algorithm-driven platforms to achieve their milestones through artificially intelligent solutions that can ease and accelerate the process of diagnosis as well as interpretations for the doctors.

In the Indian scenario, the culture of healthcare start-ups based on AI is mostly in tier 1 cities and the trend is likely to extend to tier 2 cities with the increasing number of IT professionals interested in multidimensional uses of their capabilities. Also, the effective utilization of funds and stimulating groups of venture capitalists are also increasing in numbers, some of who belong to neither health care nor software development, but are still interested in investing into such ventures. Angel investors and mentor groups associated with organizations like BIRAC which are working with the prime objective of making healthcare start-ups a success add to this ecosystem. To top all this, the 'make in India' campaign initiated by the Government of India supports this through its policies through 'Atal Incubation Mission' (AIM), where constant funds are disbursed to universities and similar institutions for the development of incubation centers, where new start-ups can be incubated without having to worry about space, instrumentation, and basic facilities. National eHealth Authority (NeHA) was proposed for the development of an integrated health information system by the Ministry of Health and Family Welfare in 2015 in India. It will be the nodal authority that will develop an integrated health information system along with the application of telemedicine and mobile-health by collaborating with various stakeholders. The 'Task Force on AI for India's Economic Transformation' was set up by the Ministry of Commerce and Industry in 2017 to explore possibilities to leverage AI for development across various fields.

The contemporary trends of utilization of AI in healthcare services remain in day-to-day monitoring of vital data points, diagnostics, and predictive analytics followed by deep learning (Table 5). Companies like Tricog (Bangalore), Lybrate (Faridabad), LiveHealth (Pune), Practo (Bangalore), MUrgency (Mumbai), Portea (Bangalore), Advancells (Noida), Forus (Bangalore), AddressHealth (Bangalore), Mitra Biotech (Bangalore) are strongly surfacing serving the fields of m-Health, e-Health, medical emergency services in one app, stem cell therapy, blindness, pediatrics, and numerous other dimensions [34].

Table 4 Companies dedicated to the medical data-mining sector [32, 33]

Products/companies	Founding year	Features	Location
(1) Zephyr Health	2007	• The start-up combines databases, machine learning algorithms, and as its biggest plus, great data visualization to help healthcare companies gain insight into a diverse set of data more quickly • The start-up was selected as one of the 2016s 100 Most inspiring people in the life sciences industry by PharmaVOICE magazine readers	Newark, California, USA
(2) Google Deepmind Health	2010	• Predetermined algorithms are used to mine medical records in order to provide better and faster health services • This also includes experimenting with disease monitoring technologies – Collaborators—Moorfields Eye Hospital NHS Foundation Trust – Verily, the life sciences arm of Google's umbrella corporation – Alphabet is working on its genetic data-collecting initiative, the baseline study	London, UK
(3) Sentrian	2012	• Their fundamental objective is to eliminate all preventable hospital admissions through remote patient monitoring • Health-related predictions for individuals on the basis of their database • Firstly, it collects the vital data points in details noting even the subtle signs through biosensors • And then, it interprets and predicts through its machine learning technology	Aliso Viejo, California, USA

(continued)

Table 4 (continued)

Products/companies	Founding year	Features	Location
(4) CareSkore	2014	• A real-time, cloud-based predictive analytics platform • CareSkore basically predicts through its Zeus algorithm in, based on combination of clinical, laboratories, demographic and behavioral data, how likely a patient will be readmitted to a hospital • Hospitals improve the quality of care based on their input, while patients could also get a clearer picture about their health • Special communication service platform is available for patients with notifications about their risks and issues	Chicago, USA
(5) CloudMedX Health	2014	• Another data-mining start-up situated in the heart of the Silicon Valley • Focuses on optimizing patient and financial outcomes through predictive analytics	Palo Alto, California, USA
(6) Oncora Medical	2014	• Focus area—cancer research and treatment, especially in radiation therapy • They have built a data analytics platform that can help doctors design sound radiation treatment plans for patients • Seed fund of $1.2 million received in 2016 • In 2017, it will be extending its services of precision radiation oncology platform to three major medical centers to help 10,000 patients receive personalized treatments	Philadelphia, Pennsylvania, USA

(continued)

Table 4 (continued)

Products/companies	Founding year	Features	Location
(7) IBM WatsonPaths	2015	• It consists of two cognitive computing technologies • They help physicians make more informed and accurate decisions faster and to cull new insights into electronic medical records (EMR) • Collaborator—Cleveland Clinic Lerner College of Medicine of Case Western Reserve University	New York, USA

In the global context, AI already holds a big share of market in the healthcare sector. There are devices that can automatically detect diseases like anemia, malaria, leukemia, and other cancers, thus reducing waiting time of the patients to get pathology reports before treatment. There are also a category of devices, which are used to continuously monitor normal health parameters like blood pressure, heart rate, and sugar level (Table 2). Noninvasive techniques like 'liquid biopsy' (spit test) cut off all the pains associated with the diagnosis of cancer and act as first line of treatment as even a day saved in diagnosis is a chance won to survive the disease. The general methods of peeping inside the living systems like cardiac imaging, MRI, ECG, ultrasonography, microscopy, tissue analysis, radiology are known and are evolving at a pace faster, which is another level of research altogether. This ever-growing and unrolling world of technologies in diagnostics and analysis, to unfold the inner secrets of human body, when amalgamated to data collection, deep learning, and AI, become a powerful combat against disease and illness.

The process of drug discovery is a long tedious and exorbitant process. A single over the counter (OTC) drug may also prove extortionate for the pharmaceutical business sometimes. Here, AI coalesced with bioinformatics (Table 3) seems to be the game changer by working out years' job in hours, plus the saved resources are an advantage. This, however, is directly not related to patients, as here the customers are usually drug manufacturing companies, but the saved resources would definitely lead to reduced prices of medicines and treatment for patients.

Another aspect of AI-powered devices is that they reduce complete dependence on medically trained staff. Some devices are so user friendly and portable that the sophistication and complexity of a typical hospital-type environment are completely kept at bay. Where on one hand devices like portable dialysis machine (Table 1) give the patient the ease of mobility, the devices like absorbable heart stent eradicate the botheration of aftereffects. Smart pill like medication today makes it completely possible to track the exact time of ingestion of drug, and devices like speech restorer and autonomous wheelchair not only act as augmented body parts but also impact the patient's personal and social life by heightening their self-confidence.

Table 5 Companies based in India striving in the field of artificial intelligence-assisted healthcare services [36]

Products/company	Year	Features	CEO
i. Advenio Technosys Location—Chandigarh	2010	• Machine learning-based computer-assisted detection (CADx) • These services are applicable in those clinical settings where resources are less and burden is more • They are into diagnostics of respiratory infections through researching algorithms and analyzing images	Mausumi Acharyya
ii. Aindra Location—Bangalore	2012	• Focus area—cancer diagnostics, especially cervical cancer • Digital microscopy is followed by pathologist's confirmation, which is automatically learned by machine	Adarsh Natarajan
iii. Ten3T Location—Bangalore	2014	• Focus area—cardiac care remote monitoring system • Cicer is Ten3T's first wearable medical device that fits into an adult's palm • It is loaded with a biomedical sensor coupled with machine learning methods • This sensor streams continuous data in real time to the doctors • It monitors real-time ECG, respiration, pulse, temperature • It being tested for multiple healthcare settings like in private practice, nursing homes, or for at-home monitoring	Sudhir Borgonha
iv. Qorql Location—Noida	2014	• An online healthcare start-up that uses AI and big data to enhance the productivity of doctors and to help patients manage their ailments	Sanjay Singh
v. SigTuple Location—Bangalore	2015	• Product name—Shonit • AI pathologist for medical diagnosis with reduced time and effort • Analysis of medical images, scans, and videos to generate information for diagnosis • Automated procedure to detect diseases like anemia, malaria, leukemia, and other diseases reduced waiting time • A smartphone and a mechanical component are attached to regular microscope for imaging samples • Product name—Manthana. An AI-driven continuous learning platform that provides solutions for automated analysis of peripheral blood smear, urine and semen sample, retinal scans, and chest X-rays	Rohit Pandey

(continued)

Table 5 (continued)

Products/company	Year	Features	CEO
vi. VectorDoc Location—Bangalore	2015	• They are developing a mobile app for preliminary diagnosis with high accuracy • It allows real-time collaboration between healthcare providers, nurses, pharmacists, laboratories, and hospitals • They aim to deliver health care in resource-constrained areas also like in rural and poor urban areas	Thekkethala Pyloth Vincent and Joffy Vincent
vii. Touchkin Location—Bangalore	2015	• It uses smartphone sensors which can sense changes in the patterns of communication, activity, and sleep • Touchkin platform collects the data from mobile phones and sensors and uses the data to identify changes in behavioral patterns	Jo Aggarwal
viii. HealthMir Location—New Delhi	2015	• The app developed by the company provides health-related information through videos • It facilitates personalized health trackers	Abhash Kumar
ix. Niramai Health Analytix Location—Bangalore	2016	• Cancer diagnostics, especially breast cancer solutions • A thermal analytics-based pain-free, portable, light, and small screening device • It utilizes big data analytics, artificial intelligence, and machine learning for reliable, early, and accurate breast cancer screening	Geetha Manjunath
x. Qure.ai Location—Mumbai	2016	• Qure.ai is a decision support tool for diagnosis of diseases from radiology and pathology • The start-up offers personalized cancer treatment plans from histopathology imaging and genome sequences • Deep learning assists the machines to learn from various sources and to interpret medical images quickly and accurately to generate useful data	Prashant Warier
xi. Predible Health Location—Bangalore	2016	• Medical imaging targeting advanced knowledge in the cancer treatment powered by a neutral cloud computation platform • It focuses on building radiology solutions for improving accuracy and efficiency of radiologists and physicians	Suthirth Vaidya
xii. Orbuculum Location—Chennai	2017	• Predictive interpretation of genomic data in a fast, cost-effective, and accurate way • Diseases such as cancer, diabetes, neurological and cardiovascular disorders can be predicted early	Pranav Gangwal

The most apparent application of AI is patient data recording and mining (Table 4). Using a robot to explain laboratory results to patients or taking a chatbot as a life coach is no more an exaggeration now. The number of efficient workers in health care is not a big number already, and this goes down further as some members are busy only to maintain patient data records. Here, AI has played a vital role in helping the medical world with data collection, maintenance, and deriving a meaningful interpretation through deep learning. This data mining is helpful not only for doctors and medical staff, but these days insurance companies also use this information to digitally verify patient's insurance information to ensure it is valid and accurate and digitization reduces the manual load. With the growing population, it is not possible to proportionally increase the number of health practitioners per person, but it is possible to make the process easier and faster through AI.

The interpretation based on data size is sometimes more than even the number of cases handled by the health professionals, but anyways AI is used as a helping hand and not as an alternative for doctors. Artificial intelligence has also accelerated technology-assisted programs in areas like mental health (emotional intelligence indicators and post-traumatic stress disorder treatment program is already in market), rehabilitation and dentistry. Healthcare research is as vast as the curiosity of human mind, and the challenges are ever changing with changing times. This is of course just a beginning of wireless medical technologies, and a lot of research still needs to be done on at actual application of such devices including the psychological responses of the subject. Moreover, this paradigm shift in medical culture also needs to be supported and endorsed by doctors as common people still consider a doctor's advice over gadgets, and if so, the extended expenditures should be justified. As nature finds out ways to balance human population through constantly evolving diseases, human quest to conquer death, decay, and injury continues with all the more preparedness.

Acknowledgements I would like to thank my mentor Dr. A. K. Saxena and Dr. Ram Vishwakarma for their visionary sensibilities in the field of science. Also, I would like to acknowledge my BIRAC experience, as they funded a start-up Nurture Earth R&D Pvt. Ltd, where I worked for 2 years. This exposure was the mind-set changer, which molded the researcher inside me into an entrepreneur.

Conflicts of Interest The author declares no conflict of interest.

References

1. Zara, S.: Everything you need to know about Sophia. The world's first robot citizen. Forbes (2017). Available from https://www.forbes.com/sites/zarastone/2017/11/07/everything-you-need-to-know-about-sophia-the-worlds-first-robot-citizen/#5a6887d046fa. Accessed 12 Jan 2018
2. Adam, C.E.: I spoke to the future and the future stared back at me, blankly. Gizmodo (2018). Available from https://gizmodo.com/i-spoke-to-the-future-and-the-future-stared-back-at-me-1821900011. Accessed 12 Jan 2018
3. Tandon, R.: Artificial intelligence is a game changer for personal devices: Gartner. Businessworld (2018). Available from http://businessworld.in/article/Artificial-Intelligence-Is-A-Game-Changer-For-Personal-Devices-Gartner/10-01-2018-136912/. Accessed 12 Jan 2018

4. Yan, H., Huo, H., Xu, Y., Gidlund, M.: Wireless sensor network based E-health system-implementation and experimental results. IEEE Trans. Consum. Electron. **56**(4), 2288–2295 (2010). https://doi.org/10.1109/tce.2010.5681102
5. Hernandez, A.I., Mora, F., Villegas, M., Passariello, G., Carrault, G.: Real-time ECG transmission via Internet for nonclinical applications. IEEE Trans. Inf. Technol. Biomed. **5**(3), 253–257 (2001). https://doi.org/10.1109/4233.945297
6. Ball and Lillis: E-health: transforming the physician/patient relationship. Int. J. Med. Informatics **61**(i1), 1–10 (2001)
7. Hsu, J., Huang, J., Kinsman, J., Fireman, B., Miller, R., Selby, J., Ortiz, E.: Use of e-health services between 1999 and 2002: a growing digital divide. J. Am. Med. Inform. Assoc. **12**(i2), 164–171 (2005)
8. Istepanian, R.S.H., Emil, Jovanov, Zhang, Y.T.: Guest editorial introduction to the special section on m-health: beyond seamless mobility and global wireless health-care connectivity. IEEE Trans. Inf. Technol. Biomed. **8**(4), 405–414 (2004)
9. Bhanumurthy, B.M.: What lessons can we learn from the first wave of AI? World economic forum (2017). Available from https://www.weforum.org/agenda/2017/01/what-did-we-learn-from-the-first-wave-of-ai/. Accessed 12 Jan 2018
10. Feldman, M.: China puts forth plan to become world leader in artificial intelligence. Top 500 (2017). Available from https://www.top500.org/news/china-puts-forth-plan-to-become-world-leader-in-ai/. Accessed 12 Jan 2018
11. CB Insights Research Briefs: The United States of artificial intelligence startups (2017). Available from https://www.cbinsights.com/research/artificial-intelligence-startup-us-map/. Accessed 12 Jan 2018
12. Tractica: Artificial intelligence revenue to reach $36.8 billion worldwide by 2025 (2016). Available from https://www.tractica.com/newsroom/press-releases/artificial-intelligence-revenue-to-reach-36-8-billion-worldwide-by-2025/. Accessed 12 Jan 2018
13. Tractica: Healthcare artificial intelligence software, hardware, and services market to reach $19.3 billion worldwide by 2025 (2017). Available from https://www.tractica.com/newsroom/press-releases/healthcare-artificial-intelligence-software-hardware-and-services-market-to-reach-19-3-billion-worldwide-by-2025/. Accessed 12 Jan 2018
14. Pitt, E.: 3D printing breakthrough lands grad top job. University of Wollongong Australia (2013). Available from-https://media.uow.edu.au/news/UOW145874.html. Accessed 13 Jan 2018
15. Pitt, E.: BioPen to rewrite orthopaedic implants surgery. University of Wollongong Australia (2013). Available from https://media.uow.edu.au/news/UOW162803.html. Accessed 13 Jan 2018
16. Bhunia, P.: 3D printing, robotic surgery to create custom implants for bone cancer patients in Australian research project. Opengov (2017). Available from https://www.opengovasia.com/articles/3d-printing-robotic-surgery-to-create-custom-implants-for-bone-cancer-patients-in-australian-research-project. Accessed 13 Jan 2018
17. Pareek, S.: Indian scientists develop a portable and affordable diagnostic kit for a blood cell disorder. The better India (2014). Available from https://www.thebetterindia.com/13558/indian-scientists-develop-portable-affordable-diagnostic-kit-blood-cell-disorder-sickle-cell-anaemia/. Accessed 13 Jan 2018
18. Pan, Shaowu, Yang, Zhibin, Chen, Peining, Deng, Jue, Li, Houpu, Peng, Huisheng: Wearable solar cells by stacking textile electrodes. Angew. Chem. Int. Ed. **53**(24), 6110–6114 (2014). https://doi.org/10.1002/anie.201402561
19. Yang, Zhibin, Deng, Jue, Sun, Xuemei, Li, Houpu, Pen, Huisheng: Stretchable, wearable dye-sensitized solar cells. Adv. Mater. **26**(17), 2643–2647 (2014). https://doi.org/10.1002/adma.201400152
20. Coosemans, J., Hermans, B., Puers, R.: Integrating wireless ECG monitoring in textiles. Sens. Actuators A Phys. **130–131**, 48–53 (2006)
21. Linz, T., Gourmelon, L., Langereis, G.: Contactless EMG sensors embroidered onto textile. In: Proceedings of the 4th International Workshop on Wearable and Implantable Body Sensor Networks, Aachen, Germany, vol. 13, pp. 29–34 (2007)

22. Löfhede, J., Seoane, F., Thordstein, M.: Soft textile electrodes for EEG monitoring. In: Proceedings of 2010 the 10th IEEE International Conference on Information Technology and Applications in Biomedicine (ITAB), Corfu, Greece pp. 1–4 (2010)
23. Löfhede, J., Seoane, F., Thordstein, M.: Textile electrodes for EEG recording—a pilot study. Sensors **12**, 16907–16919 (2012)
24. Bella, T.: Who needs shoes? These socks are 15 times stronger than steel. Men'shealth (2017). Available from https://www.menshealth.com/fitness/socks-stronger-than-steel. Accessed 13 Jan 2018
25. Wenner, M.: New biotech breakthroughs that will change medicine. Popular Mechanics (2009). Available from http://www.popularmechanics.com/science/health/a4180/4303407/. Accessed 17 Dec 2017
26. Business Wire: Xcorporeal, Inc. Announces the XCR-6 dialysis platform for self-directed kidney hemodialysis (2008). Available from http://www.businesswire.com/news/home/20081112005484/en/Xcorporeal-Announces-XCR-6-Dialysis-Platform-Self-Directed-Kidney. Accessed 17 Dec 2017
27. Fornell, D.: FDA approves first totally Bioresorbable Stent. DAIC (2016). Available from https://www.dicardiology.com/product/fda-approves-first-totally-bioresorbable-stent. Accessed 17 Dec 2017
28. Heater, B.: Taking a ride in MIT's self-driving wheelchair. Techcrunch (2017). Available from https://techcrunch.com/2017/06/24/taking-a-ride-in-mits-self-driving-wheelchair/. Accessed 17 Dec 2017
29. Paul, M.: Promising new nanotechnology for spinal injury. Northwestern (2008). Available from http://www.northwestern.edu/newscenter/stories/2008/04/SpinalCordInjury.html. Accessed 17 Dec 2017
30. Proteus Digital Health: Otsuka and Proteus announce the First U.S. FDA approval of a digital medicine system: Abilify MyCite (aripiprazole tablets with sensor) (2017). Available from https://www.proteus.com/press-releases/otsuka-and-proteus-announce-the-first-us-fda-approval-of-a-digital-medicine-system-abilify-mycite/. Accessed 17 Dec 2017
31. Cision PR Newswire: Arterys Inc. Arterys partners with GE healthcare to launch transformative cardiac medical imaging platform (2015). Available from-https://www.prnewswire.com/news-releases/arterys-partners-with-ge-healthcare-to-launch-transformative-cardiac-medical-imaging-platform-300188657.html. Accessed 17 Dec 2017
32. The Medical Futurist: Top artificial intelligence companies in healthcare to keep an eye on (2017). Available from http://medicalfuturist.com/top-artificial-intelligence-companies-in-healthcare/. Accessed 22 Apr 2018
33. Flagella, D.: Machine learning healthcare applications-2018 and beyond. Tech emergence (2018). Available from https://www.techemergence.com/machine-learning-healthcare-applications/. Accessed 25 Apr 2018
34. Roy, S.: These 10 startups are redefining healthcare in India. Your story-Healthcare (2018). Available from https://yourstory.com/2018/01/healthtech-startups-india/?. Accessed 1 May 2018
35. Wang, X., Kaczor-Urbanowicz, K.E., Wong, D.T.W.: Salivary biomarkers in cancer detection. Med. Oncol. **34**(1), 7 (2017). https://doi.org/10.1007/s12032-016-0863-4
36. Saha, A.T.: Artificial intelligence based healthcare startups in India. Medgenera (2017). Available from https://medgenera.com/news/12-artificial-intelligence-healthcare-startups-india-ai/. Accessed 25 Apr 2018

Point of Care Devices

Tabassum Zafar

Abstract In the recent digital era, biotechnology has also expanded the wings in the development of sophisticated intelligent devices. Points of care (POC) devices are one of the most recent and sophisticated digitalized applications of biotechnology and biochemistry. POC devices are a result of collaboration between biotechnology, biochemistry, electronics and marketing. POC devices employ technology for detection of various diseases and disorders without centralized, expensive, laboratory-specific equipment and tools. POC device is one of the best examples of lab-on-a-chip advancement, where only a small POC chip can efficiently perform the work in absence of a whole laboratory set-up. The best about POC devices is their easy, convenient, automated operational procedures, which made them user-friendly. Nowadays, where users prefer customized less time-consuming, portable and reliable methods for various purposes, POC devices are the most suitable technology to meet the demands. Nowadays, POC devices are serving various sectors of society, including healthcare, food safety and personal care. Major applications of POC devices are the test kits available in the market for detection of many clinical conditions. Pregnancy and diabetes test kits are the most popular and relatively cheaper older examples of POC technology. POC testing is a rapidly growing application of biotechnology, which offers new advancements and enormous growth possibilities for biotech professionals. Purpose of the present chapter is to focus on the attention of readers towards this relatively new, less popular but commercial application of biotechnology.

Keywords Point of care devices · Lab on a chip · Ancillary testing
Intelligent clinical devices

T. Zafar (✉)
Department of Biosciences, Barkatullah University, Hoshangabad Road, Bhopal 462026, Madhya Pradesh, India
e-mail: tztabassumzafar@gmail.com

M. Khoobchandani and A. Saxena (eds.), *Biotechnology Products in Everyday Life*, EcoProduction, https://doi.org/10.1007/978-3-319-92399-4_5

1 Introduction

Over the years, biotechnology has extended contributions in social convenience beyond the conventional approach. In the similar relevance, intelligent devices are one of the remarkable milestones in the field of healthcare management. Traditional clinical care is now having an alternate, improved and accurate testing tool named Point of care (POC) testing. POC devices are minimized assay devices, which are used for detection of any clinical condition or health issue by immediate diagnostic testing at the site where clinical care is delivered to patients [1]. POC testing is also known by various other names including near-patient testing, patient-focused testing, bed-site testing, satellite testing, ancillary diagnosis, alternate site testing, lab-on-a-chip testing and peripheral testing [2]. POC devices are modern diagnostic tools to perform rapid, accurate, cost-efficient, portable, on-site diagnostic testing without any involvement of complicated instrumentation to facilitate continence.

In the twenty-first century, various advancements are available in the field of diagnostics. Microfluidics is a branch of science, which is based on the fluid characteristics in the microenvironment. POC devices are one of the best applications of microfluidics. Recent advancements in the field of microfluidic sciences are the basics behind most of the POC systems.

Availability of POC testing is a revolutionary boon in the sector of health management that facilitates improved and immediate clinical assistance, when patients are supposed to get tested on their bed site. In contrast to classical testing procedures, POC devices are portable with the possibility of an immediate shift in response to the need of the user. The limitation of confined space laboratory testing is overcome by the availability of point of care devices. Rapid diagnosis, reduced cost, high throughput, convenience, portability are some of the major facts, which prove the suitability of these devices over conventional laboratory techniques [3].

Most of the chips and devices work on the principles of immunology and biochemistry. Development of any new point of care system is a vigorous and lengthy process, which requires lots of background research with specific range of standardizations. Optimization and standardization of lab-on-a-chip technique require highly sophisticated, sterile and ideal set-up, but once the miniature identification techniques related to any disease have been developed by researchers, the cost of peripheral testing turns relatively lower than the conventional procedures. Mass production of POC devices for population actually reduces the cost by hundreds of times for patients.

The purpose of the present chapter is to provide an overview of the POC devices in interestingly and concise manner to make the young readers aware of the most popular technological advancement in the field of microfluidics. The present chapter summarizes a critical, unbiased overview of POC industry and also briefly discusses the recent advancements, future perspectives, along with the limitations of current trends.

2 Biochemistry Behind the POC Testing

POC devices interact with various classes of analytes, including proteins, nucleic acids, cells, enzymes, immunological markers, electrolytes and many other small molecules [4]. Once presence of these analytes is detected by POC intelligent device, immediately samples are allowed to react with the pre-existing substrates in order to perform a reaction, which results in the formation of the respective product. Product formation is usually identified by the end user either qualitatively in the form of color change or qualitatively in the form of signal. Sometimes it is required to amplify the low signal to enhance the test precision. Protein-based clinical specimens such as whole blood, serum, plasma, saliva, urine, and few other samples are clinically tested by both immunoassays and enzymatic microdevice-based POC assay [5].

Some examples of protein-based ancillary tests include detection of viral infections (anti-HIV antibodies, antibodies against influenza A/B virus, rotavirus antigens), bacterial infections (antibodies against Streptococcus A and B, *Chlamydia trachomatis*, *Treponema pallidum*), parasitic infections (histidine-rich protein 2 for *P. falciparum*, trichomonas antigens), and non-communicable diseases (PSA for prostate cancer, C-reactive protein for inflammation, HbA1c for plasma glucose concentration). Saliva-based nano-biochip immunoassay is another popular example of protein analyte-based POC testing, which has been constructed to detect a panel of C-reactive protein, myoglobin, and myeloperoxidase in acute myocardial infarction patients [4, 6–8].

Cell-based POC devices are generally used for the hematological purposes. These methods do not employ the use of any proteins or biologies to achieve cell separation. Size-dependent, density-based techniques and affinity-based methods are usually incorporated in this category of POC device, e.g. CD4C lymphocyte counting used to monitor the progression of HIV/AIDS is one of the classic examples of cell-based POC testing [9].

POC devices are capable to perform highly sensitive, fully integrated molecular diagnostic testing with remarkable efficacy. Nucleic acid testing (NAT) is based on the identification of organism-specific DNA or RNA sequences. Nucleic acid detection and analysis is the key to identify the source of infection, pathogen and disease. In the similar connection, HIV (diagnosis and viral load monitoring), H_1N_1 influenza, tuberculosis, and group B streptococcal disease testing POC devices are some noteworthy examples [10]. Electrolytes are also utilized in some cases for ancillary testing. Many POC devices are sensitive enough to identify the status of sodium, potassium, chlorine, calcium as electrolytes in the sample. They can also monitor the amount of blood gases using pH change, potentiometric, amperometric and conductance measurements. Color-dependent pH analysis of blood is one of the widely used electrolyte-based POC microfluidic devices [4, 11, 12].

3 Remarkable Features of POC Devices

POC devices have a narrow range of universally accepted properties. In order to meet the user's need, an identical POC device is supposed to carry some remarkable feature as mentioned below [13].

- POC devices should be user-friendly.
- POC devices should be affordable.
- POC devices should be sensitive enough to minimize the possibility of false-negative or false-positive results.
- POC devices should be rapid and robust in storage and use.
- POC devices should provide results should be concordant with an established laboratory method.
- Results of POC testing should be delivered to the end-user.

4 Clinical Applications of POC Testing

The applied methods, area and utility of POC devices are very vast and overlapping. Although POC test systems are initially meant to be used by medical professionals to provide immediate assistance to patients within their immediate vicinity, nowadays POC devices are widely used for self-clinical assessment. However, a test performed without a medical indication cannot be strictly described as POC testing, but blood glucose monitoring devices and pregnancy test kits are some of the popular examples of self-monitoring POC devices. These devices also frequently used in hospitals and healthcare centres. Some classic examples of POC testing, which brought revolution in the healthcare industry, include assessment of lipid markers, hematological assessment, C-reactive protein assay, serum bilirubin assessment, determination of hepatic and cardiac injury markers. Investigation of more than one parameters sequentially or simultaneously with automatic calibration of hand-held microfluidics and microsensor-based POC devices have become very convenient [11, 14, 15]. Blood gas analyser (BGA) is a ubiquitous tool to determine level of the blood gases, acid-base equilibrium and electrolyte concentrations. Blood gas analyzer and blood sugar analyzers are long-established POC systems with the capabilities to process samples for the assessment of more than 100 parameters [16]. Point of care determination of blood sugar employs a small, mobile instrument which works on enzymatic reactions, followed by photometric or electrochemical detection [17].

Individual clinical biochemistry parameters with a wide analytical spectrum such as creatinine, urea, glutamate oxalate transaminase (GOT) and glutamate pyruvate transaminase (GPT) tests are also performed by POC devices nowadays. POC determination of C-reactive protein (CRP) is used in hospitals and practices for the rapid evaluation of the necessity of antibiotic therapy. Pharmacy-based testing of lipid parameters is another POC prototype [18]. Detection of hyperbilirubinemia in neonates is a popular example of using immediate photometric serum bilirubin

POC testing [19]. The popularity of POC device is evident by enormous utility of these intelligent devices to determine quick values of various markers such as cardiovascular injury markers, e.g. immunological determination of cardiac troponins, BNP (brain natriuretic peptide), D-dimers (fibrin degradation product, FDP) along with hemostaseological assessment including international normalized ratio (INR), partial thromboplastin time (PTT) and activated clotting time (ACT) in whole blood [20–22].

The outbursting utility is not only confined in these spaces, but it is extending its branches in various critical areas, including detection of HIV, malaria, dengue, sexually transmitted infections, various blood infections [23–31]. Nucleic acid detection is also an application of nanotechnology-based point of care devices [32, 33]. The field of lab-on-a-chip testing is not only limited to classic bed-site testing but also now various advancements are available according to the necessity and economy. Little recent advancement in intelligent device industry includes contact lens glucose sensors, tattoo-based sensors and smart holograms [34–36].

5 Advantages of POC Devices Over Existing Technologies

POC testing devices broadly organized into two categories, namely professional category and non-professional category. Professional category usually deals with the emergency, critical care and extensive care ancillary testing related to infectious diseases, cardiac markers, diabetes, lipid and hematology. The non-professional term is used for 'over the counter' point of care products including glucose monitoring and pregnancy detection kits. Diabetes testing is the largely used non-professional point of care device followed by pregnancy testing while peripheral testing of infectious diseases is the fastest growing professional point of care zone [37].

New advancements in point of care patient management made the clinical care very portable and organized. The advantages of POC devices over other existing technologies include drastically reduced assay volume, fast diagnostic processing, and quick readouts. POC devices are very convenient as compared to high-efficacy cutting-edge technologies. In comparison to POC devices, next-generation sequencing, droplet-based microfluidics and many such techniques are time-consuming and difficult. Some of the advantages of peripheral testing are discussed below.

Cost efficiency Cost-efficacy is one of the remarkable properties of POC devices in comparison to classical instrumentation earlier required for the diagnostic purpose. Micro-POC testing systems are a relative cost-efficient option as it reduces the cost of sample transportation, maintenance of laboratory set-up and management of costly chemical stock. Many satellite-testing devices are meant for single-time immediate, use and therefore, they do not employ any additional maintenance cost [38].

Quick readability Clinical laboratory testing required a technical procedure including various steps like the recommendation of test on prescription, laboratory visit or sample collection, manual or instrument-based experiment and preparation of reports. This time-consuming procedure is not suitable for the cases where immediate assistance is required. POC devices are the most suitable tool for such cases because of its quick readability and immediate results.

Portability Classical laboratory techniques required incorporation of huge sophisticated immobile instrumentations organized in wide physical space. POC testing overcomes the major limitation of physical laboratory set-up and provides a portable alternative. A POC device is like the lab on a chip, where the test performed at clinical laboratory could easily perform by using a small chip like mini device. Portability is a remarkable feature of peripheral-testing devices, which make it a suitable option for testing at the remote area, testing at the site of any natural disaster or bed site during emergencies. Portability of these devices reduces the transport time and provides immediate results during emergencies [39].

Automation Automation is the basic the key feature of point of care diagnosis. POC devices detect clinical conditions in a fully automated manner with least involvement of manual laboratory operating protocol. Automatic bed-site testing facilitates the rapid and accurate clinical testing. Automation not only reduces the involvement of manpower required for the conventional testing method but also minimizes the possibilities of man-made errors [1, 39].

Rapid processing POC systems are very fast in sample processing. In comparison to traditional laboratory methods, usually, POC testing does not require any extended incubation or long processing. When the sample is loaded on the detection site of the peripheral-testing device, it gets process immediately and provides a suitable signal to identify the completion of biochemical reaction within the device. The lack of delay in sample processing contributes in the accuracy of clinical testing in various ways. During routine laboratory-based diagnostic methods, long-term storage, incubation, temperature maintenance, transport, additions of preservatives or anticoagulant could possibly interfere with the quality of the sample. The strong possibility of activity change and loss exists in this case, while during the point of care testing immediate assessment of sample reduces such possibilities considerably [40].

Minimized sample requirement Classical instrument-based clinical assays require a certain amount of sample in order to achieve accuracy. Separation of serum for clinical tests required a little bit more blood sample comparatively, while POC testing is usually designed to perform with direct biological sample. Peripheral-testing devices are miniature-testing tools; hence, they are very beneficial for the cases where less amount of sample is available for bed-site testing [41].

Product specificity POC testing methods are very much specific to the particular test for which it is commercially prescribed to minimize the possibilities of false-positive or false-negative results [42].

6 Market Opportunities and Technology Trends

Developments of such technical advancements open up a wide area for professional possibilities in the field of research and clinical product marketing for young dynamic and enthusiastic biotech professionals. POC testing is an emerging technology that looks promising for the future. Many well-established POC technologies and improvements therein have taken place in many related areas so as to make the devices easier to use, less prone to errors, more compact and portable, sound analytical performance, which is sufficient for clinical purposes. It is likely that such incremental improvements with existing technologies will continue as the spin-off from the continual miniaturization of electronics and increased computing power. Various documentations are available to support the POC devices over existing centralized laboratory diagnosis [1, 2, 40, 43].

POC devices are equally growing their business in both developed and developing world. In 2011, POC testing devices share a remarkable market of approximately $15 billion in total $51 billion markets of the total in vitro diagnostics, which signifies the fast pace commercialized growth of these systems [39, 44]. This commercial share is expected to increase with 4% annual growth rate. The USA, Europe and Asia cover the top consumers of the point of care devices. Various marketing reports predict that global POC market is expected to reach USD 36.96 billion by 2021, at CAGR of 9.8% from 2016 to 2021, which will further build a business of $48.2 billion by 2025 worldwide [45, 46].

A sharp rise is still expected in advancements and popularization of peripheral-testing systems within few next years in Western healthcare markets. However, meeting the demand for POC devices at a reasonably low price is still a challenging task for health-care industry.

7 Discussion

POC is, essentially, the delivery of care wherever the patient may be and a POC technology is defined as a technology that is used for bringing care to the patient rather than taking the patient to care. POC intelligent devices are one of the revolutionary biochemical and biotechnological and biochemical advancement in the field of clinical health management to facilitate less fragmented, patient centerd diagnosis [47].

The lab-on-a-chip concept is an emerging clinical testing tool, which will extend enormous possibilities of research and employment in the coming few decades. These are one of the best widely popular commercial products of biotechnology and biochemical diagnosis. The development of POC devices becomes possible with the collaboration of various interdisciplinary streams, including biotechnology, biochemistry, immunology, electronics and marketing in focused, ethical and professional way. Most of the lab-on-a-chip systems work on biochemical and immunological principles. Some devices are supposed to provide the quantitative results, while some others are capable to quantify the sample. These various functions decide the various principles employed for kits development. The popularization of point of care devices is increasing with each passing year by health professionals and patients. Realizing the full potential of POC technologies represents a critical factor in advancing overall healthcare sector by making the predictive, pre-emptive, preventive and personalized care of the future accessible to all communities [48].

Every coin has two sides; similarly, the POC device industry is also facing various complexities beyond the existing advancement. POC testing may sometimes exhibit considerable differences in analytical sensitivity and precision in compare to the routine laboratory testing. One of the basic limitations of POC devices is that these tests are largely restricted to automatically measurable parameters. Like other in vitro diagnostic tests, POC testing devices also have limitations related to legal conditions such as certification, licensing, operation and quality assurance. Reducing the cost of point of care devices for the mass availability is another big challenge for the healthcare professionals [49].

Numerous POC testing devices are available in the market with different protocols and standardization. In clinical healthcare, POC devices employ according to the need of the patient, availability of different devices, financial limitations and ethical standards. These possibilities may affect the selection of appropriate tool for immediate use. Despite its portability and apparent simplicity, POC testing systems also face many, analytical and post-analytical issues like any other centralized laboratory. An abrupt procedure, unsuitable storage, improper handling may affect the accuracy of point of care devices [50].

Maintaining high-level accuracy coordinated with operator technical competency is a rising challenge for healthcare industry. To overcome the existing challenges, an interdisciplinary healthcare system is desired with technical expertise to overcome the present challenges of POC intelligent device industry.

POC is a rapidly growing application of biotechnology, which offers new advancements and enormous growth possibilities as a researcher and entrepreneur. The future prospect of portable intelligent device-based healthcare industry is very bright and glorious. Advancements in the lab-on-a-chip technique are supposed to create fully automated clinical diagnostic to replace centralized laboratories. Involvement of enthusiastic, dynamic professionals is required for the POC healthcare industry to contribute in the field of research; policy-making and marketing to achieve fast pace commercialization and globalization.

Acknowledgements The author extends her sincere acknowledgement to Dr. Basheerulla Shaik, Assistant Professor, National Institute of Technical Teachers Training and Research, Bhopal, India for, his motivational support and constructive criticism.

Conflicts of Interest The author declares no conflict of interest.

References

1. Witters, D., Sun, B., Begolo, S., Rodriguez-Manzano, J., Robles, W., Ismagilov, R.F.: Digital biology and chemistry. Lab Chip **14**(17), 3225–3232 (2014)
2. Nichols, J.H.: Management of near-patient glucose testing. Endocrinol. Metab. Serv. Training Continuing Educ. **12**, 325–34 (1994)
3. Yetisen, A.K., Akram, M.S., Lowe, C.R.: Paper-based microfuidic point-of-care diagnostic devices. Lab Chip **13**, 2210–2251 (2013). https://doi.org/10.1039/C3LC50169H
4. Chin, C.D., Linder, V., Sia, S.K.: Lab-on-a-chip devices for global health: past studies and future opportunities. Lab Chip **7**, 41–57 (2007). https://doi.org/10.1039/b611455e
5. Hall, D.A., Ptacek, J., Snyder, M.: Protein microarray technology. Mech. Ageing Dev. **128**, 161–167 (2007). https://doi.org/10.1016/j.mad.2006.11.021
6. Floriano, P.N., Christodoulides, N., Miller, C.S., Ebersole, J.L., Spertus, J., Rose, B.G., Kinane, D.F., Novak, M.J., Steinhubl, S., Acosta, S., Mohanty, S., Dharshan, P., Yeh, C.K., Redding, S., Furmaga, W., McDevitt, J.T.: Use of saliva-based nano-biochip tests for acute myocardial infarction at the point of care: a feasibility study. Clin. Chem. **55**, 1530–1538 (2009). https://doi.org/10.1373/clinchem.2008.117713
7. Von Lode, P.: Point-of-care immuno testing: Approaching the analytical performance of central laboratory methods. Clin. Biochem. **38**, 591–606 (2005). https://doi.org/10.1016/j.clinbiochem.2005.03.008
8. Madhivanan, P., Krupp, K., Hardin, J., Karat, C., Klausner, J.D., Reingold, A.L.: Simple and inexpensive point-of-care tests improve diagnosis of vaginal infections in resource constrained settings. Trop. Med. Int. Health **14**, 703–708 (2009). https://doi.org/10.1111/j.1365-3156.2009.02274.x
9. Radisic, M., Iyer, R.K., Murthy, S.K.: Micro- and nanotechnology in cell separation. Int. J. Nanomed. **1**, 3–14 (2006)
10. Niemz, A., Ferguson, T.M., Boyle, D.S.: Point-of-care nucleic acid testing for infectious diseases. Trends Biotechnol. **29**, 240–250 (2011). https://doi.org/10.1016/j.tibtech.2011.01.007
11. Baum, J.M., Monhaut, N.M., Parker, D.R., Price, C.P.: Improving the quality of self-monitoring blood glucose measurement: a study in reducing calibration errors. Diabetes Technol. Ther. **8**, 347–357 (2006)
12. Boyd, J.C., Bruns, D.E.: Effects of measurement frequency on analytical quality required for glucose measurements in intensive care units: assessments by simulation models. Clin. Chem. **60**, 644–650 (2014). https://doi.org/10.1373/clinchem.2013.216366
13. Pearce, D.M., Shenton, D.P., Holden, J., Gaydos, C.A.: Evaluation of a novel electrochemical detection method for Chlamydia trachomatis: application for point-of-care diagnostics. IEEE Trans. Biomed. Eng. **58**, 755–758 (2011). https://doi.org/10.1109/TBME.2010.2095851
14. Price, C.P., Koller, P.U.: A multicentre study of the new reflotron system for the measurement of urea, glucose, triacylglycerols, cholesterol, gamma-glutamyltransferase and haemoglobin. J. Clin. Chem. Clin. Biochem. **26**, 233–250 (1988)
15. Van Solinge, W.W., Huisman, A.: Point-of-care testing in haematology. In: Price, C.P., St. John, A., Hicks, J.M. (eds.) Point-of-Care Testing, 2nd edn. AACC Press, Washington, DC, pp. 415–20 (2004). ISBN-13: 978-1594250125
16. Walters, M.C., Abelson, H.T.: Interpretation of the complete blood count. Pediatr. Clin. North Am. **43**, 599–622 (1996)
17. Kristensen, G.B.B., Christensen, N.G., Thue, G., Sandberg, S.: Between-lot variation of external quality assessment of glucose: clinical importance and effect on participant performance evaluation. Clin. Chem. **51**, 1632–1636 (2005). https://doi.org/10.1373/clinchem.2005.049080
18. Dahler-Eriksen, B.S., Lauritzen, T., Lassen, J.F., Lund, E.D., Brandslund, I.: Near-patient test for C-reactive protein in general practice: assessment of clinical, organizational, and economic outcomes. Clin. Chem. **45**(4), 478–485 (1999)
19. Thaler, M., Luppa, P., Schlebusch, H.: die bilirubinbestimmung – Eine aktuelle Übersicht. J. Lab. Med. **32**, 1–10 (2008). https://doi.org/10.1515/JLM.2008.005

20. Nola, K.M., Gourley, D.R., Portner, T.S., Gourley, G.K., Solomon, D.K., Elam, M., Regal, B.: Clinical and humanistic outcomes of a lipid management program in the community pharmacy setting. J. Am. Pharm. Assoc. **40**, 166–173 (2000)
21. The joint European society of Cardiology/American college of cardiology committee: Myocardial infarction redefined—A consensus document of The Joint European society of Cardiology/American college of Cardiology committee for the redefinition of myocardial infarction. Eur. Heart. J. **21**, 1502–1513 (2000)
22. Spannagl, M., Calatzis, A.: Point-of-care analyse des Hämostasesystems. J. Lab. Med. **26**, 68–76 (2002)
23. Dunning, L., Hsiao, N.Y., Myer, L.: Point-of-care HIV early infant diagnosis: is test sensitivity everything? J. Int. AIDS Soc. **18**, 20235 (2015). https://doi.org/10.7448/ias.18.1.20235. eCollection 2015
24. Kim, S., Nhem, S., Dourng, D., Ménard, D.: Malaria rapid diagnostic test as point-of-care test: study protocol for evaluating the VIKIA® Malaria Ag Pf/Pan. Malar J. **14**, 114 (2015). https://doi.org/10.1186/s12936-015-0633-3
25. Gan, V.C., Tan, L.K., Lye, D.C., Pok, K.Y., Mok, S.Q., Chua, R.C.R.: Diagnosing dengue at the point-of-care: utility of a rapid combined diagnostic kit in Singapore. PLoS ONE **9**, e90037 (2014). https://doi.org/10.1371/journal.pone.0090037
26. Pearce, D.M., Styles, D.N., Hardick, J.P., Gaydos, C.A.: A new rapid molecular point-of-care assay for Trichomonas vaginalis: preliminary performance data. Sex Transm. Infect. **89**, 495–497 (2013). https://doi.org/10.1136/sextrans-2012-051000
27. Gaydos, C.A., Van Der Pol, B., Jett-Goheen, M., Barnes, M., Quinn, N., Clark, C., Daniel, G.E., Dixon, P.B., Hook, E.W., Lark, C.: CT/NG study group. Performance of the cepheid CT/NG expert rapid pcr test for detection of chlamydia trachomatis and neisseria gonorrhoeae. J. Clin. Microbiol. **51**, 1666–1672 (2013). https://doi.org/10.1128/JCM.03461-12
28. Vassall, A., Van Kampen, S., Sohn, H., Michael, J.S., John, K.R., Den Boon, S., Davis, J.L., Whitelaw, A., Nikol, M.P., Gler, M.T., Khaliqov, A., Zamudio, C., Perkins, M.D., Boehme, C.C., Cobelens, F.: Rapid diagnosis of tuberculosis with the Xpert MTB/RIF assay in high burden countries: a cost-effectiveness analysis. PLoS Med. **8**, e1001120 (2011). https://doi.org/10.1371/journal.pmed.1001120
29. Stimpe, F., Karathanos, A., Droppa, M., Metzger, J., Rath, D., Müller, K., Tavlaki, E., Schäffeler, E., Winter, S., Schwab, M., Gawaz, M., Geisler, T.: Impact of point-of-care testing for CYP2C19 on platelet inhibition in patients with acute coronary syndrome and early dual antiplatelet therapy in the emergency setting. Thromb. Res. **134**, 105–110 (2014). https://doi.org/10.1016/j.thromres.2014.05.006
30. Ritzi-Lehnert, M.: Development of chip-compatible sample preparation for diagnosis of infectious diseases. Expert Rev. Mol. Diagn. **12**, 189–206 (2012). https://doi.org/10.1586/erm.11.98
31. Spencer, D.H., Sellenriek, P., Burnham, C.A.: Validation and implementation of the GeneXpert MRSA/SA blood culture assay in a pediatric setting. Am. J. Clin. Pathol. **136**, 690–694 (2011). https://doi.org/10.1309/AJCP07UGYOKBVVNC
32. Raja, S., Ching, J., Xi, L., Hughes, S.J., Chang, R., Wong, W., McMillan, W., Gooding, W.E., McCarty Jr., K.S., Chestney, M., Luketich, J.D., Godfrey, T.E.: Technology for automated, rapid, and quantitative PCR or reverse transcription-PCR clinical testing. Clin. Chem. **51**, 882–890 (2005)
33. Hartman, M.R., Ruiz, R.C., Hamada, S., Xu, C., Yancey, K.G., Yu, Y., Han, W., Luo, D.: Point-of-care nucleic acid detection using nanotechnology. Nanoscale **5**, 10141–10154 (2013). https://doi.org/10.1039/c3nr04015a
34. How Google's smart contact lens works. Sydney Morning Herald. http://www.smh.com.au/technology/sci-tech/how-googles-smart-contact-lens-works-20140121-317q4.html (2014). Last accessed 13 Oct 2017
35. Bandodkar, A.J., Hung, V.W., Jia, W., Valdés-Ramírez, G., Windmiller, J.R., Martinez, A.G., Ramirez, J., Chan, G., Kerman, K., Wang, J.: Tattoo-based potentiometric ion-selective sensors for epidermal pH monitoring. Analyst **138**, 123–128 (2013). https://doi.org/10.1039/c2an36422k. Epub 2012 Oct 31

36. Kabilan, S., Marshall, A.J., Horgan, A., Creasey, C.D., Kew, S.J., Dean, K.E.S., et al.: "Smart" holograms—a novel diagnostic platform. Nanotechnology **3**, 467–470 (2006)
37. St John, A., Price, C.P.: Economic evidence and point-of- care testing. Clin. Biochem. Rev. **34**, 61–74 (2013)
38. Laurence, C.O., Moss, J.R., Briggs, N.E., Beilby, J.J.: POCT; POCT Trial management group. The cost-effectiveness of point of care testing in a general practice setting: results from a randomised controlled trial. BMC Health Serv. Res. **10**, 165 (2010). https://doi.org/10.1186/1472-6963-10-165
39. Price, C.P., Kricha, L.J.: Improving health care accessibility through point-of-care technologies. Clin. Chem. **53**, 1665–1675 (2007). https://doi.org/10.1373/clinchem.2006.084707
40. St John, A., Price, C.P.: Existing and emerging technologies for point-of-care testing. Clin. Biochem. Rev. **35**(3), 155 (2014)
41. Junker, R., Schlebusch, H., Luppa, P.B.: Point-of-care testing in hospitals and primary care. Dtsch. Arztebl. Int. **107**, 561–567 (2010). https://doi.org/10.3238/arztebl.2010.0561
42. Kulinsky, L., Noroozi, Z., Madou, M.: Present technology and future trends in point-of-care microfluidic diagnostics. Methods Mol. Biol. **949**, 3–23 (2013). https://doi.org/10.1007/978-1-62703-134-9_1
43. Scientia advisors. Strategic review of point-of-care diagnostics. Available from: http://www.scientiaadv.com/data.cfm [Google Scholar]
44. Scientia advisors. Strategic review of the molecular diagnostics market. Cambridge, MA, USA (2013)
45. http://www.marketsandmarkets.com/PressReleases/point-of-care-diagnostic.asp. Accessed 13 Oct 2017
46. http://www.prnewswire.com/news-releases/482-billion-point-of-care-diagnostics-market-analysis–trends-report-2016–2025—growing-government-support-to-increase-the-adoption-of-poc-devices—research-and-markets-300244149.html. Accessed 13 Oct 2017
47. Issadore, D., Westervelt, R.M. (eds.): Point of care diagnostics on a chip. In: Biological and Medical Physics, Biomedical Engineering. Springer, Heidelburg NewYork, London (2013). https://doi.org/10.1007/978-3-642-29268
48. Boncheva, M., Pascaleva, I., Dineva, D.: Performance of POCT chemistry analyzer Picollo (Abaxis) in primary health care. Gen. Med. **4**, 28–31 (2002)
49. Clinical and laboratory standards institute (CLSI). Point-of-care blood glucose testing in acute and chronic care facilities. Approved guideline, 3rd edn. CLSI POCT12-A3. Wayne, PA, USA (2013)
50. Kumar, S., Kumar, S., Ali, M.A., Anand, P., Agrawal, V.V., John, R., Maji, S., Malhotra, B.D.: Microfluidic integrated biosensors: prospects for point-of-care diagnostics. Biotechnol. J. **8**, 1267–1279 (2013). https://doi.org/10.1002/biot.201200386

Cardiac Pacemaker—A Smart Device

Xuwei Hou

Abstract Maintaining a normal heart rate (between 60 and 100 beats per min) and rhythm (beats regularly) is vital for human health. The significant reduction in heartbeat (less than 60 beats per min) is called bradycardia. It has enhanced mortality rate because of the lack of effective treatment. Treatment patterns are changing, since 1960s a pacemaker was used for the first time in a clinical setting and revolutionized the bradycardia therapy. In this chapter, readers will learn the history of modern pacemaker, mechanism, function, and its pros and cons. Readers will be surprised to discover how smart a modern pacemaker is. Till now, pacemaker is the definitive treatment for a drug refractory bradycardia and has saved tens of thousands of lives worldwide.

Keywords Pacemaker · Bradycardia · Cardiomyopathy · Cardiovascular disorder

1 Introduction

Pacemaker is a small electronic device that is implanted in the chest of the patients to pace heart. What makes a modern pacemaker distinguished from other thousands of medical devices is that it is a smart device, which means a pacemaker can track and sense patient's need for heartbeat and adjust to the best working mode to fit that need. It works automatically in a small self-powered chamber similar to a cookie in size, without any external intervention. More importantly, pacemakers seldom cause error during their whole working lifespan. Hence, a modern pacemaker can fully replace our natural pacing cells in the heart to trigger heart muscle contraction. Amazingly, a modern pacemaker can do something our natural heart cannot do. Pacemakers improve pumping function in a heart with weak muscles and resynchronize the heartbeat in an obstructive cardiomyopathy.

X. Hou (✉)
School of Medicine, University of Missouri, Columbia, MO 65211, USA
e-mail: beckhxw007@gmail.com

M. Khoobchandani and A. Saxena (eds.), *Biotechnology Products in Everyday Life*, EcoProduction, https://doi.org/10.1007/978-3-319-92399-4_6

2 Basic Knowledge About Our Heart and Cardiac Arrhythmia

To begin with, let's understand basic knowledge about our heart. Our heart is composed of four chambers, namely the left and right atria, and left and right ventricles (Fig. 1).

Cardiac muscle contraction is triggered by electrical impulses generated at a sinoatrial node located at the upper part of right atria of the heart. The sinoatrial node consists of a cluster of cells (called pacing cells) that can automatically generate regular electrical pulses at the frequency between 60 and 100 per min. The electrical pulses are like bioelectricity. They are very week in intensity, but enough to trigger left and right ventricles to contract almost at the same time (because right ventricle beats little behind its left counterpart, it is normal). The sinoatrial node is the heart's natural pacemaker.

The heart also has its own conduct system, made of special fiber tissues, to spread electrical signals throughout the heart rapidly and evenly causing the heart muscle to contract in a coordinated, rhythmic pattern (Fig. 1). As a result, the blood is ejected effectively from heart to whole body. It should be noted that, in the conjunction area between atria and ventricle, there is another important node called atrio-ventricular node. The electrical signals will have a brief delay (0.12–0.20 ms, shorter than human blink eyes), followed by rapid conduction downward into ventricular muscles.

3 Cardiac Arrhythmia

It is an abnormality in heartbeating. It can be either the rate or the rhythm of heartbeat, which means that the heart beats either too quickly (>100 beat per min, called

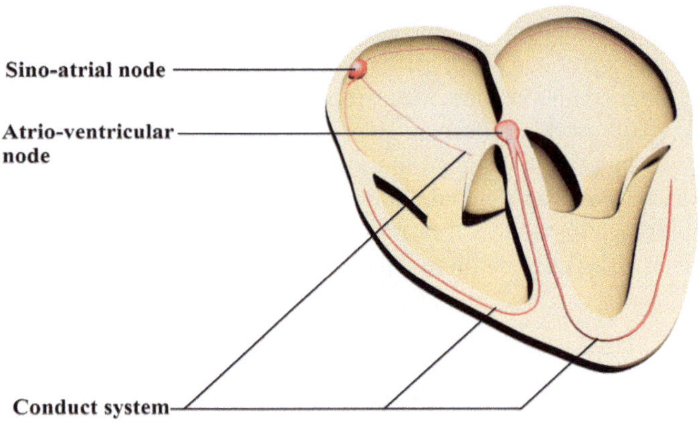

Fig. 1 The pacing and conduct system in heart (Image created by Xuwei Hou)

tachycardia), too slowly (<60 beat per min, called bradycardia), or irregularly (such as atrial fibrillation). As such, the electric impulses cannot be generated or conducted properly, causing a shortage in a heartbeat. A patient will have different symptoms such as palpitations, dizziness, black vision, and shortness of breath, chest discomfort, exercise intolerance, fainting, or even a seizure onset, depending on the severity of cardiac arrhythmia. They can cause fatal consequence if they are not properly treated. Cardiac arrhythmia has enhanced mortality rate in the past, of which a major portion of death was occurred from severe bradycardia. Pacemaker was initially invented to treat severe bradycardia, and it functions perfectly in replacing sinoatrial node, and/or, atrio-ventricular node. Thousands and thousands of lives have been saved since its clinical use in the 1960s.

4 The History of Pacemaker

Prior to the advent of pacemakers, doctors have no reliable treatment for patients suffering from bradycardia, and many died of severe heartbeat loss. No medication, including atropine, epinephrine, can maintain patient's heart rate for a long time.

The first pacing machines in human history came into being between late 1920s and early 1930s [1]. The Australian anesthesiologist Mark Lidwell and the American physiologist Albert Hyman shared the credit for the first external cardiac pacemaker. They worked independently, and they developed their own cardiac pacing machines. Lidwell's device needed a needle to be placed into the patient's ventricle. In 1928, he used his machine to generate electrical impulses and saved the life of a child born with cardiac arrest. Hyman is the person who invented the name "pacemaker" for his pacing-aid device (Fig. 2) in 1932 [2, 3]. He experienced lots of frustrations due to technical problems in pacemaker and faced many oppositions, including that of the Journal of the American Medical Association. As a result, his device had never been used clinically.

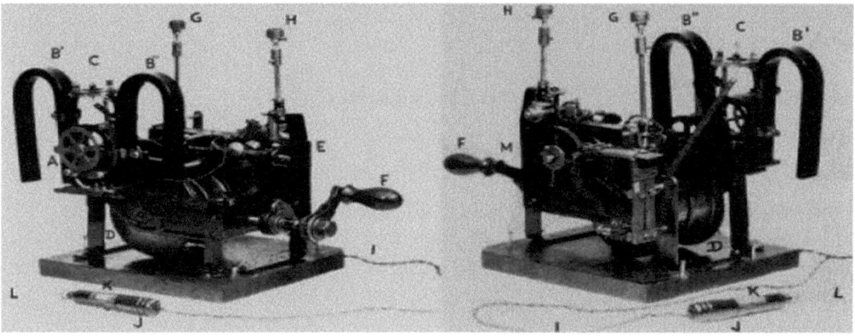

Fig. 2 Albert Hyman's "pacemaker". *Source* https://www.ncbi.nlm.nih.gov/pmc/articles/PMC3232561/

The pace of heart pacing technology accelerated in early 1950s [4, 5]. A Canada researcher Wilfred Bigelow restored heartbeat to a dog with hypothermia induced a cardiac arrest, using a device generating electrical impulses. In 1951, a Boston cardiologist, Dr. Paul Zoll designed a tabletop pacemaker (Fig. 3) that was successfully used for the treatment of heart block, which means one or several electrical impulses are lost during their conduction, so no impulse stimulates ventricle to contract [6]. Dr. Zoll saved several lives during his practice using this device. He published the first study on the use of external pacing for cardiac arrest treatment [7, 8]. However, this unit being so bulky and heavy that it had to be carried on a cart. In 1956, two British researchers, Aubrey Leatham and Geoffrey Davies, introduced an external simulator with a demand capability. That is a big progress in pacing technology because it allows the pacemaker to sense the intrinsic cardiac activities. Pacemaker started to get smart at this time [9].

Late 1950s and early 1960s are the "Golden Years" in pacemaker history. These ten years witnessed many important achievements in the domain of cardiac pacing technology. In 1957, Earl E. Bakken, an American electrical engineer and a TV repairman, produced the first battery-operated wearable pacemaker (Fig. 4). This is another a leap in the history of pacemaker, because it makes the in vivo use possible. Bakken later established a company called Medtronic Inc., which now is a giant in the pacemaker industry. The first pacemaker implantation was performed on October 8th, 1958, in Sweden. The pacing system (Fig. 5) had been developed by the surgeon Ake Senning, and the physician inventor Rune Elmqvist implanted on a 43-year-old engineer, Arne Larsson. He had been hospitalized because of 20–30 heart arrests daily despite maximized medication treatment. His condition was hopelessly poor. In desperation, his wife learned from a local newspaper that Dr. Rune Elmqvist in Stockholm developed an implantable pacemaker. Under her insist, Mr Larsson was transferred to Stockholm and had the implantation procedure, and he survived [10]. Because of the short life of battery, he had to go back to hospital to replace a new one every six months. As the first human to receive an implanted pacemaker, he had been implanted a total of 26 different pacing devices during his life. He died in 2001 at the age of 86. It was pacemakers, which extended his life from 1958 to 2001, a total of 43 years of life!

5 Components of a Modern Pacemaker

A modern pacemaker usually consists of a lithium battery, an electrical impulse generator, also called stimulator, and wires with sensors at their tips (also called leads, or electrodes). The battery supplies power to the whole system, which is housed in a thin metal box with excellent biocompatibility. The wires connect the generator to the heart.

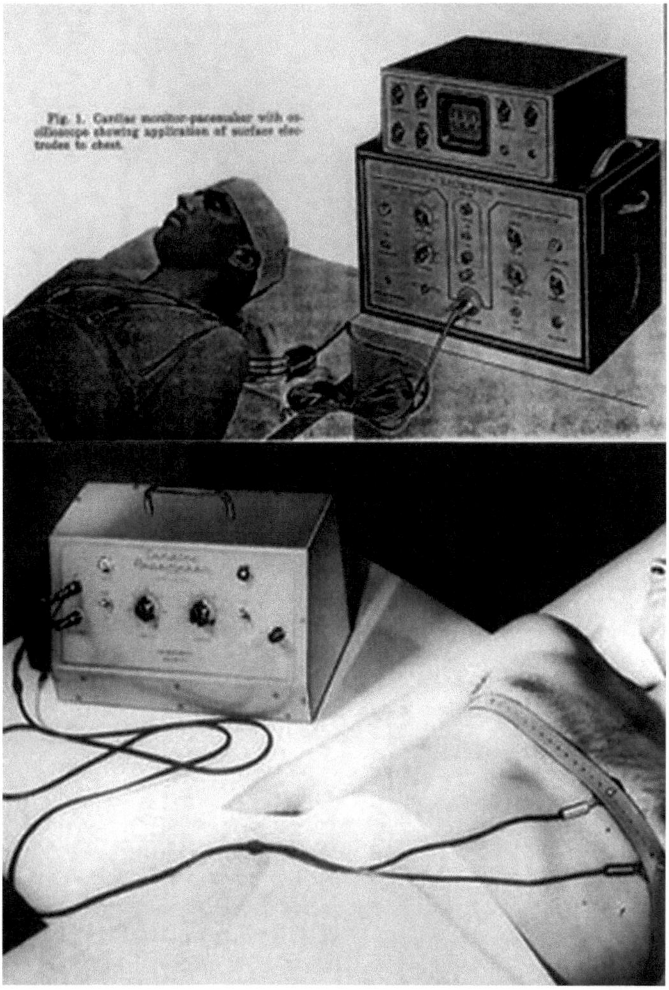

Fig. 3 Zoll's external pacemaker

6 Basic Functions of Modern Pacemaker

Two basic functions, called sensing and pacing, are seen in all modern pacemakers. Modern pacemakers can continuously monitor the intrinsic cardiac activity of patient heart though its sensors built on the end of wires. Normal heart beats and the pacemaker will sense that will not interrupt our natural heart activity, which lessens power consumption from battery. This is called sensing function. Once the sensor on top of the pacemaker lead detects a slow heart rate or interrupted conduction, either from a problem of sinus node or conduct system, the stimulator of pacemaker will generate an electrical impulse to replace our natural heart and start new heart-

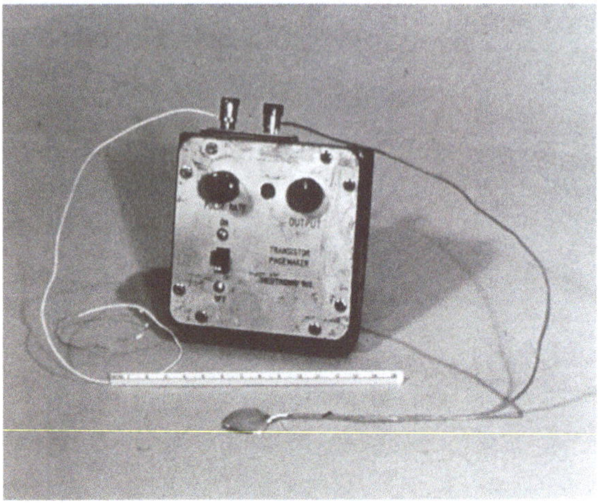

Fig. 4 Bakken's pacemaker with leads. *Source* https://www.ncbi.nlm.nih.gov/pmc/articles/PMC3232561/

Fig. 5 First-implanted pacemaker. *Source* https://www.ncbi.nlm.nih.gov/pmc/articles/PMC3232561/

beats. This is termed as pacing function. Besides, the basic functions, all modern pacemakers have an advanced feature called "rate-responsive capability function." This function depends on a more sophisticated sensor installed on top of pacemaker wires that can detect the patient's physical activity, respiratory rate, blood temperature, blood carbon dioxide level, and other physiological parameters, and switch to the best pacing mode to fit the patient's physiologic need. All these steps happen automatically and accurately.

7 Temporary and Permanent Pacemaker

Based on the duration of usage, pacemaker can be classified as temporary and permanent. Temporary pacemakers are designed for short-term use only. They are only used when the arrhythmia arises due to a transient or emergent cause, or, to protect patients until a permanent pacemaker is placed. The generator of a temporary pacemaker is placed outside patient's body, and taped to the skin or attached to a belt or to the patient's bed. A medical staff will continuously monitor the patients. Medical staff will perform regular examinations to monitor for any possible complications. Temporary pacemakers will be only used in hospitals or clinics, and will be eventually removed, or replaced by a permanent pacemaker.

In contrast, permanent pacemakers are for long-term used to treat arrhythmia, which lasts long, or causes severe clinical outcome. These pacemakers are always located inside the patient's body. Patients will live with a permanent pacemaker for long time for the rest of their life.

8 Pacing Modes

After 60 years of development, various pacemakers and pacing modes have been developed and used to restore and sustain a regular heartbeat. Based on the pacing sites, pacemakers are classified as single, dual, or triple chambered. The pacing site for single-chamber pacemakers is either the right atrium or right ventricle, which is usually used for patients with single node disease, either sinoatrial node, or atrio-ventricular node. The pacing mode of single-chamber pacemakers can be further classified as single-chamber atrial pacing and single-ventricle pacing [11–14].

In single-chamber atrial pacing mode, the lead of pacemaker is permanently placed in the right atrium. The candidates for pacing mode are those with sinoatrial node dysfunction only, which means their atrio-ventricular node still function. This is a problem of electrical pulses production, but not their conduction. The pacemaker will track normal atrial activity, and remains silent if the natural pacing cells work normally. Once pacemaker detects the atrial activity is under normal limit value, e.g., lower than 60 beats per min, it will generate electrical pulses to replace the failed natural pacing cells. In the case of single ventricular pacing mode, the lead is placed in the right ventricle, which is used in patients with atrio-ventricular node dysfunction. Under this mode, pacemaker will generate electrical pulses and send these pacing signals directly to the apex of right ventricle, from where the pacing signal will spread to the rest of the heart. This mode is enough to save life during a cardiac emergency; however, it sometimes causes a troublesome problem called pacemaker syndrome, a phenomenon associated with conflict between intrinsic cardiac activity of a patient and pacemaker signals, causing loss of atrium and ventricles synchrony in beating. Patient will have symptoms such as palpitation, dizziness, and short of breath [15–17]. If that happens, upgrade to dual-chamber system will be the only definite way to treat pacemaker syndrome [18].

Dual-chamber pacemakers have two leads, which stimulate right atrium and right ventricle [19]. This mode is used for patients with both sinoatrial node and atrioventricular node dysfunctions. Dual-chamber pacemaker allows a heart rhythm to be more naturally and acts like the normal activities of the heart [20]. That is the reason why this mode is also called "physiological pacing." As a result, "pacemaker syndrome" seldom occurs under this mode. This pacing mode is widely used in patients worldwide, but the cost is considerably higher than single-chamber pacing [21, 22].

Dual-chamber pacing is an advanced pacing mode. The ventricle will be paced following every sensed atrial event, up to a preset maximum rate. If the patient develops an irregular atrial beat (e.g., atrial fibrillation, meaning atrium beats irregularly fast), the ventricle would also beat fast and irregularly, which is undesirable. To solve this problem, all dual-chamber pacemakers have a function called "mode switching," which means automatic reprogramming of a pacemaker to a mode that no longer tracks the intrinsic atrial rate, to avoid the fast heartbeating followed by abnormal atrial activity [23].

Triple-chamber pacing is an even more advanced mode, but it requires three independent leads. First lead in the right atrium, second in the right ventricle, and the third lead in the left ventricle. Triple-chamber pacing is used in patients who have dilated and weakened heart muscle (medical name is heart failure). These pacemakers "resynchronize" the left and right ventricles and improve the efficiency of the contraction of the heart [24, 26].

The biggest advantage of triple-chamber pacemakers is that they can increase heart-pumping function and improve patient's symptoms in the case of heart failure. For years, medication is the only way to treat heart failure. If patients do not response well to medications, heart transplantation is the only way to prolong their lives [27, 28]. The advent of triple-chambered pacemakers revolutionizes the heart failure treatment. Doctors can now make a patient's heartbeats better with a device. In addition to the two leads located in right atrium and right ventricle, as a dual-chamber pacemaker, a triple-chamber pacemaker has a third lead that is positioned in a vein on the outer surface of the left ventricle, which allows the pacemaker to simultaneously stimulate the left and right ventricles and restore a coordinated and synchronous pumping action. This is why it can increase the cardiac output of each heartbeat. The triple-chamber pacemaker is sometimes named as "biventricular pacemakers," or cardiac resynchronization therapy (CRT) [29].

Besides CRT, there is another pacemaker-based treatment for heart failure, called CRT defibrillators (abbreviated as CRT-D), which is a combination of pacemaker and defibrillator [30, 31]. CRT-D can rapidly terminate an abnormally fast, life-threatening heart rhythm frequently seen in a heart failure patient. If the device senses arrhythmia, it immediately delivers an electrical shock (called defibrillation) to the heart.

9 How to Implant a Pacemaker

A pacemaker implantation is typically conducted in a sterile laboratory or operating room by an experienced cardiologist. Only a local anesthesia on the chest skin is needed; so, patients remain conscious during the whole procedure. The pacemaker is inserted into soft tissue beneath the skin in an area below the clavicle. The pacemaker leads are placed into a major vein. During the procedure, doctors can check the position of these leads to make sure they are secured within the proper regions inside the heart. The typical time for this procedure is somewhere between 1 and 3 h. However, in an emergent situation, a temporary pacemaker implantation can perform on the bedside by experienced doctors to save patient's life.

After surgery, the patients need some restrictions on arm movement and limited activities for the first month to avoid the dislocation of leads. To optimize the pacemaker function, doctors can always program the implanted device from outside using an electrical device, to set some important parameters, such as baseline heart rate and the upper heart rate at which the pacemaker will pace.

10 New Advances in Pacemaker

One of the most recent innovative technologies in pacemaker industry is leadless pacemaker [32–34]. A leadless pacemaker is small, less than ten percent the size of a conventional modern pacemaker. No skin pouch is need for a leadless pacemaker since it is directly inserted into the apex of right ventricle through a special steerable catheter, making the pacemaker implantation procedure much easier to doctors and much safer to patients. The procedure-related complications, including device pocket infection and lead failure, are dramatically decreased in leadless pacemakers (Fig. 6) [35–38].

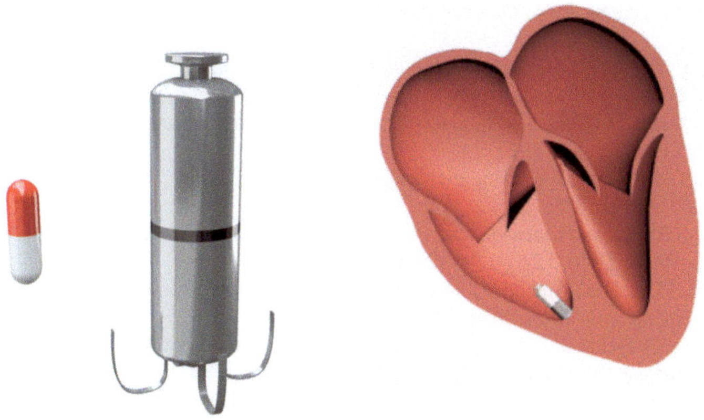

Fig. 6 A leadless pacemaker and its installation in heart (Image created by Xuwei Hou)

Acknowledgements Author thanks to Mr. Deepak Abhaya Srisrimal, Dhanvantari Nano Ayushadi Pvt Ltd, Chennai, India, for reviewing this chapter.
Conflicts of interest. The author declares no conflict of interest.

References

1. Aquilina, O.: A brief history of cardiac pacing. Images Paediatr. Cardiol. **8**(2), 17–81 (2006)
2. Sowton, E., Flores, J.: Natural history of pacemaker patients. Bull. N. Y. Acad. Med. **47**(8), 999–1010 (1971)
3. Tarjan, P.P.: History of cardiac pacing-I. Am. J. Cardiol. **41**(3), 614–615
4. Hori, M.: History of cardiac pacing and pacemaker update. Nihon Rinsho Japan. J. Clin. Med. **48**(2), 227–232 (1990)
5. Furman, S.: Early history of cardiac pacing and defibrillation. Indian Pacing Electrophysiol. J. **2**(1), 2–3 (2002)
6. Krason-Rzeszutek, E., Lelakowski, J., Haberka, K., Sedziwy, L.: The natural history of patients with an artificial pacemaker. Wiad. Lek. **41**(13), 842–845 (1988)
7. Abelmann, W.H., Paul, M.: Zoll and electrical stimulation of the human heart. Clin. Cardiol. **9**(3), 131–135 (1986)
8. Braun, P.: Paul Zoll and the discovery of pacemaker. Ther. Hung. **40**(1), 44–46 (1992)
9. Parsonnet, V.: History of cardiac pacing-II. Am. J. Cardiol. **41**(3), 615 (1978)
10. Stoney, W.S., Finger, F.E., Alford Jr., W.C., Burrus, G.R., Frist, R.A., Thomas Jr., C.S.: The natural history of long term cardiac pacing. Ann. Thorac. Surg. **23**(6), 550–554 (1977)
11. Tang, A.S., Roberts, R.S., Kerr, C., Gillis, A.M., Green, M.S., Talajic, M., Yusuf, S., Abdollah, H., Gent, M., Connolly, S.J.: Relationship between pacemaker dependency and the effect of pacing mode on cardiovascular outcomes. Circulation **103**(25), 3081–3085 (2001)
12. Montanez, A., Hennekens, C.H., Zebede, J., Lamas, G.A.: Pacemaker mode selection: the evidence from randomized trials. Pacing Clin. Electrophysiol.: PACE **26**(5), 1270–1282 (2003)
13. Ochagavia Calvo, A., Baigorri, Gonzalez F.: Selection of pacemaker stimulation mode. Med. Intensiva **30**(5), 218–222 (2006)
14. Rudzinski, T., Ciesielczyk, M., Religa, W., Zebrowski, M., Bednarkiewicz, Z., Krzeminska-Pakula, M.: Elderly and patients with sick sinus syndrome have lower chances for appropriate pacemaker mode selection, according to the Polish Cardiac Society recommendations—a single-centre retrospective analysis. Kardiologia polska **64**(4), 391–395 (2006); discussion 396
15. Ross, R.A., Kenny, R.A.: Pacemaker syndrome in older people. Age Ageing **29**(1), 13–15 (2000)
16. Bohm, A., Pinter, A., Dekany, M., Preda, I.: Ventriculoatrial cross-talk dependent pacemaker syndrome. Pacing Clin. Electrophysiol.: PACE **23**(6), 1062–1063 (2000)
17. Van Wallace, C.J.O.: Diagnosing and treating pacemaker syndrome. Crit. Care Nurse **21**(1), 24–31, 35 (2001); quiz 36-27
18. Gribbin, G.M., Parry, S.W.: Single-chamber versus dual-chamber pacemakers. N. Engl. J. Med. **339**(9), 631–632 (1998)
19. Barold, S.S.: Eccentricities of dual chamber implantable pacemakers. Pacing Clin. Electrophysiol.: PACE **23**(8), 1290–1294 (2000)
20. Slimane, M.L., Ben, Ameur Y.: Dual Chamber Card. Pacing. A multicenter study apropos of 353 pacemakers. La Tunis. Med. **80**(10), 624–627 (2002)
21. Rodriguez Chavez, L.L.: Dual chamber pacemakers: common problems and how to treat them. Archivos de cardiologia de Mexico **76**(Suppl 2), S214–220 (2006)
22. Dual-chamber pacemakers worth their cost. Health News **11**(4), 5 (2005)
23. Provenier, F., Jordaens, L., Verstraeten, T., Clement, D.L.: The "automatic mode switch" function in successive generations of minute ventilation sensing dual chamber rate responsive pacemakers. Pacing Clin. Electrophysiol.: PACE **17**(11 Pt 2), 1913–1919 (1994)

24. Furugen, A., Matsuda, N., Kasanuki, H.: Cardiac resynchronization therapy. Nihon Rinsho Jpn. J. Clin. Med. **64**(5), 941–948 (2006)
25. Gold, M.R.: Cardiac resynchronization therapy for heart failure. Curr. Cardiol. Rep. **8**(5), 314–315 (2006)
26. Sticherling, C., Schaer, B., Coenen, M., Kuhne, M., Ammann, P., Osswald, S.: Cardiac resynchronization therapy in chronic heart failure. Swiss Med. Wkly. **136**(39–40), 611–617 (2006)
27. Jolobe, O.M.: Health economics of cardiac resynchronization therapy. Br. J. Hosp. Med. **78**(10), 594 (2017)
28. Rademakers, L.M., de Boeck, B.W., Maessen, J.G., Prinzen, F.W.: Development of strategies for guiding cardiac resynchronization therapy. Heart Fail. Clin. **4**(3), 333–345 (2008)
29. Terminology and Diagnostic Criteria Committee JSoUiM. Introduction to indexes for cardiac resynchronization therapy (CRT) indication. J. Med. Ultrason. **41**(2), 261–274 (2014)
30. Chen, Z., Kotecha, T., Crichton, S., Shetty, A., Sohal, M., Arujuna, A., Kirubakaran, S., Bostock, J., Cooklin, M., O'Neill, M., Wright, M., Gill, J.S., Rinaldi, C.A.: Lower incidence of inappropriate shock therapy in patients with combined cardiac resynchronisation therapy defibrillators (CRT-D) compared with patients with non-CRT defibrillators (ICDs). Int. J. Clin. Pract. **67**(8), 733–739 (2013)
31. Freedman, R.A., Petrakian, A., Boyce, K., Haffajee, C., Val-Mejias, J.E., Oza, A.L.: Performance of dedicated versus integrated bipolar defibrillator leads with CRT-defibrillators: results from a prospective multicenter study. Pacing Clin. Electrophysiol.: PACE **32**(2), 157–165 (2009)
32. Lopes, N., Cavaco, D., Adragao, P., Carmo, P., Costa, F., Morgado, F.: Leadless pacemaker: a paradigm change. Revista portuguesa de cardiologia: orgao oficial da Sociedade Portuguesa de Cardiologia = Portuguese J. Cardiol.: Official J. Portuguese Soc. Cardiol. **35**(11), 627–628 (2016)
33. McCauley, B.D., Chu, A.F.: Leadless cardiac pacemakers: the next evolution in pacemaker technology. R. I. Med. J. **100**(11), 31–34 (2017)
34. Lopes, N., Cavaco, D., Carmo, P., Scanavacca, M.I., Adragao, P.: Leadless pacemaker: a new concept in cardiac pacing. Arquivos brasileiros de Cardiol. **107**(4), 381–382 (2016)
35. Reddy, V.Y., Miller, M.A., Knops, R.E., Neuzil, P., Defaye, P., Jung, W., Doshi, R., Castellani, M., Strickberger, A., Mead, R.H., Doppalapudi, H., Lakkireddy, D., Bennett, M., Sperzel, J.: Retrieval of the leadless cardiac pacemaker: a multicenter experience. Circ. Arrhythmia Electrophysiol. **9**(12) (2016)
36. Leier, M.: Advancements in pacemaker technology: the leadless device. Crit. Care Nurse **37**(2), 58–65 (2017)
37. Cappato, R.: Leadless pacemaker: an interesting solution coming back. J. Cardiovasc. Med. **19**(Suppl 1), e19–e20 (2018)
38. Source: https://www.ncbi.nlm.nih.gov/pmc/articles/PMC3232561/

Part III
Behind the Scenes Genes

CAR-T Cell Immune Therapy: Engineering T Cells to Treat Cancer

Sohinee Bhattacharyya and Anindit Mukherjee

Abstract Immunotherapy is a combination of therapeutic strategies that make use of a patient's immune system to confront tumors. Over the past few years, immunotherapy has evolved as a crucial therapeutic program to manage some types of cancer. Our immune system is a collection of cells and tissues which perform the specialized function of defending our body against foreign pathogens. Although our immune system is designed to combat foreign agents that pose a threat, it often fails to detect and eradicate tumor cells because tumor cells have evolved complex genetic adaptive mechanisms that help them evade the immune surveillance program. In order to overcome this obstacle and in order to improve tumor cell detection and eradication by the immune system, researchers have developed a novel therapeutic treatment strategy called chimeric antigen receptor T cell therapy (CAR-T cell therapy). T cells are immune cells that play a key role in shielding our body against foreign pathogens by mounting a potent immune response that ultimately eradicates abnormal cells. In brief, CAR-T cell therapy involves the isolation of T cells from patient blood, genetically altering the T cells to express receptors on their surface that specifically can bind to antigens present on the tumor cell, proliferating them in the laboratory to make millions of cells and finally, injecting back into the patient. T cells genetically manipulated in this manner harbor the capacity of multiplying inside the patient's body and can also eradicate tumor cells that express surface antigens specifically recognized by the engineered receptor. It took researchers all around the globe several years of research to have CAR-T cell therapy approved in August 2017 for the treatment of children with acute lymphoblastic leukemia (ALL). So far, CAR-T cell therapy has shown extremely promising results in patient populations, especially for patients who have developed chemo-resistance. The scope of this chapter is to provide a broad and

S. Bhattacharyya (✉)
Department of Cell, Development and Cancer Biology, Oregon Health and Science University, Portland, USA
e-mail: bhattacs@ohsu.edu

A. Mukherjee
Department of Nephrology and Hypertension, Oregon Health and Science University, Portland, USA
e-mail: mukherja@ohsu.edu

M. Khoobchandani and A. Saxena (eds.), *Biotechnology Products in Everyday Life*, EcoProduction, https://doi.org/10.1007/978-3-319-92399-4_7

general idea about all the different aspects of CAR-T cell therapy that are essential for the application of this therapeutic strategy to combat cancer. Some of these key features include selection of a specific tumor antigen to target, developing a strong co-stimulatory signaling strategy to elicit a potent immune response, enhancing the migratory properties of CAR-T cells so that they can localize to tumor sites, and finally, engineering T cells in a manner so that they can effectively proliferate and eliminate tumor cells. In conclusion, CAR-T cells hold tremendous potential for the treatment of cancer and have shown promising results in clinical trials.

1 Introduction

The concept of CAR-T cells was initially developed over 15 years ago at the Weizmann Institute in Israel by Zelig Eshhar and colleagues [1]. They conceptualized a novel strategy of genetically engineering T cells. By genetically manipulating T cells, the idea was to enable them to express tumor-antigen-specific receptors on their surface that would specifically and selectively bind to antigens present on a tumor cell, mount a potent immune response, and finally eradicate tumor cells by causing cytotoxicity. Generally, T cell receptors can only recognize antigens that are displayed in concert with major histocompatibility complex (MHC) proteins on the target cell. Cancer cells have evolved mechanisms of evading the immune surveillance mechanism by downregulating the surface expression of MHC proteins [2]. CAR-T cells have a tremendous advantage in overcoming this challenge as they have the capacity of directly binding to the protein antigen without the need of MHC presentation by the tumor cell.

2 Domain Organization of CAR

CAR consists of the following key structural components [3]:

(a) An extracellular antigen-recognition domain coupled to an extracellular spacer domain

The most commonly utilized antigen-recognition domain in CARS comprises heavy- and light-chain antibody fragments purified from a whole antibody. The heavy and light chains are usually attached together with the help of a linker region. The extracellular antigen-recognition domain confers antigen-binding selectivity and is also attributed to binding tumor cell antigen in an MHC-independent manner. Other tumor-antigen-binding domains such as endothelial growth factor peptide and heregulin have also been exploited in recent technological advances to modify and enhance CAR-T cell function [4]. The extracellular antigen-binding domain is physically separated from the intramembrane region by a spacer domain that confers physical

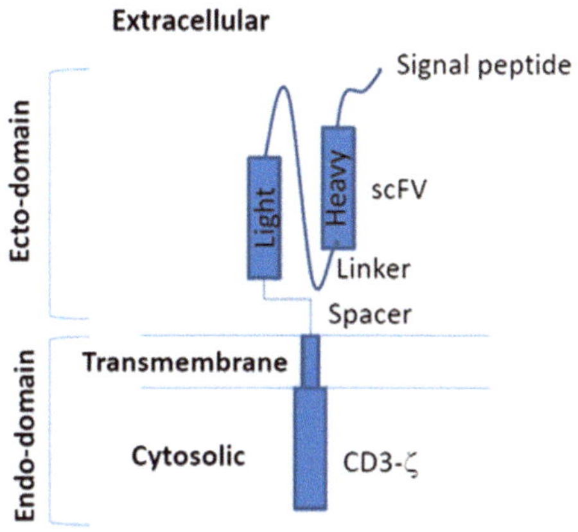

Fig. 1 Chimeric antigen receptors are a hybrid of extracellular domain or the single-chain fragment of variable region (scFV) of an antibody with the co-stimulatory domain of a T cell receptor (CD3-ς)

separation as well as added flexibility in binding to the target tumor-specific antigen. In most CAR-T cell engineering strategy, these spacer sequences are modified from antibody IgG subclasses [5] (Fig. 1).

(b) An intramembrane region

The transmembrane spanning domain aids in physically separating the extracellular domain from the intracellular signaling domain is most commonly adapted from existing sequences from T cell molecules [4, 6].

(c) Signaling/intracellular/cytosolic domain

The signaling domain is most commonly referred to as the endo-domain, and the structural engineering of this domain plays a vital role in determining the effectiveness of the immune response. There has been much debate in the field on the modeling of an optimum T cell signaling endo-domain that can potentially mount the strongest immune response without compromising on specificity. First-generation CAR-T cells utilize the CD3z chain motif [7] to mount an immune reaction. To further improve the strength of the immune response mediated by first-generation CAR-T cells, scientists engineered second-generation CARs that combined other co-stimulatory signals in addition to CD3z chain motif. One example of such a co-stimulatory signaling molecule is CD28, a co-stimulatory receptor that physically binds to B7 proteins commonly displayed on the surface of tumor cells [8]. To further improve upon this model, third-generation CARs were engineered to combine two such co-stimulatory signals to further strengthen the immune response. A few clinical studies using mice as the model system have demonstrated these third-generation CARs to have more tumor inhibitory functions and improved cytokine array production [9].

Although the basic general structure of the three generations of CAR-T cells developed so far share a large amount of overlap, clinical studies utilizing these have shown varying efficacy in clinical and preclinical studies. Some of the prime features of these three broad groups of CAR-T cells are as follows:

(a) First-generation CARs

First-generation CARs directly bind to the tumor cell-specific antigen but do not possess a co-stimulatory signal. Due to the lack of this co-stimulatory signal, they fail to mediate T cell replication, expansion, and circulation in vivo upon binding to tumor cell antigen. In summary, these first-generation CAR-T cells did not prove to be potent in eradicating tumor cells. There have been several clinical trials using first-generation CARs, but the outcomes from these studies demonstrated that this therapeutic strategy only showed modest effects in multiple different patient cohorts [7].

(b) Second-generation CARs

To overcome the weak immune responses mounted by first-generation CARs, scientists developed second-generation CARs with improved antigen-specific T cell activation. In brief, these second-generation CARs were designed to include a secondary co-stimulatory molecule such as CD28. In comparison with the first-generation CARs, second-generation CARs showed improved proliferation and cytokine in clinical studies [8, 10].

(c) Third-generation CARs

To further improve upon the immune response mounted by second-generation CAR-T cells, researchers developed third-generation CARs that include several co-stimulatory signaling motifs [9]. Although third-generation CAR-T cells showed very promising outcomes in clinical studies, one of the major pitfalls of this strategy is that such receptors may produce a very strong activation signal. Ultimately, this is a cause of great concern, as too potent an immune response can lead to potentially lethal consequences [9] (Fig. 2).

Some of the different aspects of CAR-T cell therapy that are essential for the application of this therapeutic strategy to combat cancer are summarized as follows:

- Targeting a specific antigen displayed selectively by tumor cells

Whether T cell therapy is effective or not is primarily determined by the tumor cell-specific antigen to target. If the antigen targeted is not specific to the tumor cell, the specificity of the strategy may be compromised. The most common guidelines to choosing an ideal antigen to target are chosen keeping in mind the following features:

(1) For the therapy to be safe, the antigen should be displayed only by tumor cells.
(2) The tumor cells should be dependent on the target antigen for their proliferation.

Some routinely used examples of targets for CAR-T cells include the carcinoembryonic antigen specifically expressed by colon cancer cells and folate receptor expressed on the surface of ovarian cancer cells [10, 11].

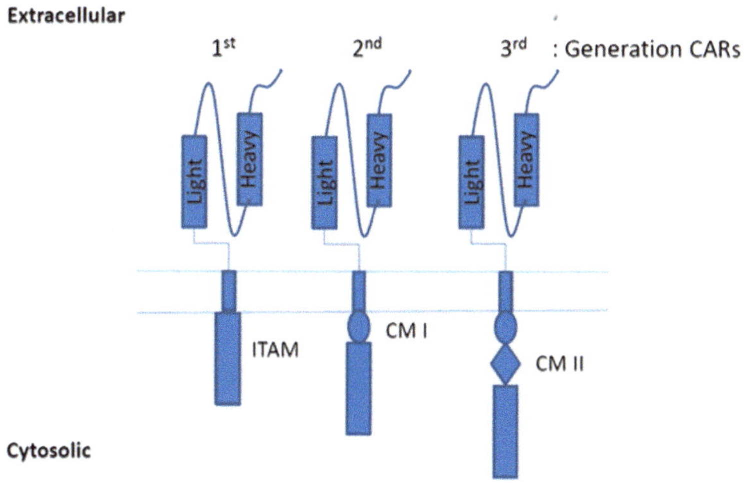

Fig. 2 Different generation of CARs: First-generation CARs have only ITAMs for signaling. Second-generation CARs have an added co-stimulatory (CM I) CD23 domain. Third-generation CARs have a second co-stimulatory domain (CM II). The co-stimulatory domains increase the effectiveness and persistence of the T cells expressing the CARs in the body

- Overcoming immune evasion by tumor cells

Tumor cells have evolved mechanisms to evade the immune response mounted against them. Tumor cells and their surrounding microenvironment have been documented in the secretion of factors such as transforming growth factor b (TGF-b) that can negatively influence the proliferation and survival of T cells in vivo. These evasion mechanisms make it extremely challenging to maintain CAR-T cells in their actively proliferating states [12]. To overcome this challenge, scientists have modified CAR-T cells to express a dominant negative (dn) TGF-b receptor utilizing retroviral transduction. These modified T cells thus become resistant to the antiproliferative effects of secreted evasive factors like TGF-b and retain their proliferative and replicative capacities in vivo [13].

- Producing an effective immune response

In order to destroy tumor cells, T cells have to survive and replicate inside the patient after infusion. Apart from proliferating in vivo, the transferred T cells have to also secrete tumor cell toxic cytokines. First-generation CAR-T cells were designed to carry out all of the above-mentioned crucial functions, but clinical and preclinical studies demonstrated poor cytokine secretion and poor replicative potential by the first-generation CAR-T cells [14–16].

In order to develop a better strategy, scientists developed second-generation CAR-T cells that in addition to the primary tumor-antigen-binding sites also incorporated a secondary stimulatory signal. One of the most frequently utilized stimulatory receptors that is expressed on the surface of second-generation CAR-T cells is the CD28

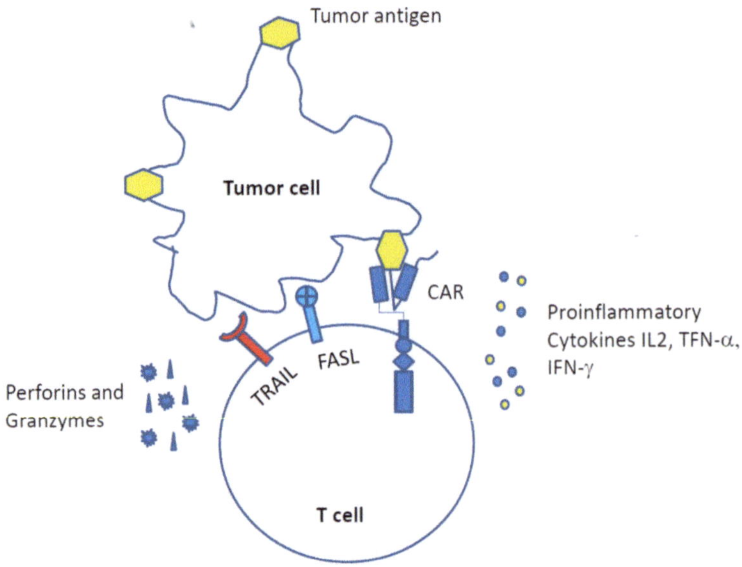

Fig. 3 Potential mechanisms of CAR-mediated toxicity: CAR expressed on the T cell recognizes and binds to tumor cell surface antigen

receptor molecule, which specifically binds with B7 tumor antigens that are displayed by a broad range of tumors [17–19]. This strategy proves to be better than the first-generation CAR-T cells in eradicating tumor cells, and several preclinical and clinical studies demonstrated increased cytokine secretion specifically that of IL-2, increased replication and survival in hosts after adoptive transfer and also over expression of proteins related to preventing spontaneous T cell death upon transfer into a patient.

To further improve on the effectiveness of second-generation CAR-T cells, researchers developed third-generation CAR-T cells that were designed to combine two stimulator signals in addition to the primary stimulatory signal. While some clinical studies have demonstrated better proliferation, cytokine production, and replication by these third-generation CAR-T cells, other clinical studies have shown the third-generation CAR-T cells to have comparable efficiency to that of second-generation CAR-T cells. One major concern with third-generation CAR-T cells is that they can mount a too strong immune response that may lead to the production of a cytokine storm inside the body of the receiving patient, ultimately leading to fatal outcomes [15, 20]. Ongoing clinical studies are currently trying to compare and test the efficacy of second- and third-generation CARs in a more in-depth and exhaustive manner (Fig. 3).

3 Trafficking and Migration of CAR-T Cells to Tumor Sites

In order to eradicate tumor cells, genetically manipulated CAR-T cells must be able to migrate to the disease sites [21]. In a clinical study conducted by Savoldo et al. [22], the trafficking of CAR-modified T cells to a skin lesion two weeks after adoptive cell transfer (ACT) was demonstrated in a patient with non-Hodgkin lymphoma (NHL) [22]. Since then, several studies have effectively demonstrated that second-generation T cells can localize to the tumor site more effectively than first-generation CAR-T cells [22–24].

- The ability of CAR-T cells to persist in vivo after adoptive transfer

For CAR-T cell therapy to be effective and successful, the infused T cells must replicate, proliferate, and survive inside the patient for an appreciable period of time. Outcomes from clinical and preclinical studies have demonstrated that in order to be successful, CAR-T cells need to survive after adoptive transfer into a patient. One such clinical study was conducted by Savoldo et al. [22]. In brief, their studies documented improved survival of T cells modified with a co-stimulatory signal containing second-generation CAR compared to first-generation CAR-T cells. Other clinical investigations have also supported these findings [23, 25]. In summary, the survival of T cells in vivo after transfer into patients is determined by the signaling motif incorporated into the CAR.

- Mounting a tumor cell cytotoxic function

Tumor cells have evolved mechanisms to evade the immune surveillance program mounted by CAR-T cells. Tumor cells can secrete several immunosuppressive secretory proteins like TGF-b that inhibit the cytolytic function of CAR-T cells in vivo [26]. In fact, researchers attribute the failures of early clinical trials using CAR-T cells to the immune evasion mechanisms exploited by tumor cells [27]. To address this major hurdle in CAR-T cell therapy, scientists have utilized multiple different strategies like preconditioning patients beforehand with chemotherapy and irradiation to reduce the number of immune evasive tumor cell populations [23] (Fig. 4).

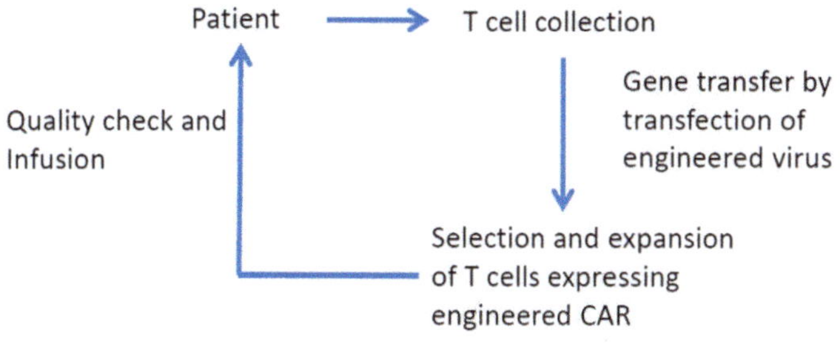

Fig. 4 A schematic overview of adoptive T cell transfer

4 Clinical Application of CAR-T Cells

4.1 Clinical Trials with First-Generation CAR-T Cells

Several centers have conducted clinical trials of first-generation CAR-modified T cells. However, first-generation CAR-modified T cells have failed to establish significant clinical benefit for antitumor efficacy [11, 27, 28].

4.2 Second-Generation CAR-Modified T Cell Trials

Based on promising preclinical data, clinical trials using T cells modified with second-generation CARs have been initiated. One study worth mentioning here is that conducted by Savoldo et al. [22]. In brief, their group treated six cancer patients by infusing them simultaneously with two different types of CAR-T cells (first-generation and second-generation CARs). In this clinical study, the results showed a better performance in all aspects in comparison with the first-generation CAR-T cells. Specifically, these studies showed the second-generation CAR-T cells to proliferate, survive, and secrete more tumor necrotic cytokines than the first-generation CAR-T cells [22]. In a separate study, researchers also demonstrated therapy with second-generation CAR-T cells resulted in partial remission of the patient's lymphoma (for up to 32 weeks) [25].

4.3 Third-Generation CAR-Modified T Cell Trial

There has been a lot of debate over the development and clinical application of third-generation CAR-T cells as they can potentially lead to lethal consequences by mounting a too strong immune response. One specific study used third-generation CAR-T cells to treat one patient with colon cancer [29]. The therapy caused a tremendous release of cytokines that lead to inflammatory responses due to toxicity from the CAR-T cells, and the patient died five days later from multi-organ failure [29].

5 CAR-Modified T Cells: Advantages

CAR-T cells can provide several advantages for cancer immunotherapy. The most prominent advantage of CAR-T cell-based immunotherapy is that it allows for the production of a large quantity of tumor-specific T cells for use in the clinical setting [30]. Secondly, CARs distinguish tumor antigens in a human leukocyte antigen (HLA)-independent manner [26]. This authorizes CAR-modified T cells to directly bind to the target antigen displayed on the surface of the tumor cell and avoids the tumor's immune evasion strategy [31]. Finally, selective and specific targeting of broad array tumor antigens like proteins, carbohydrates, and liquids is possible using CAR-T cells.

6 CAR-Modified T Cells: Major Challenges

Recent clinical studies have revealed several pitfalls in the clinical application of CAR-modified T cells that need to be addressed with the help of more exhaustive and in-depth clinical studies. One of the primary concerns regarding the application of CAR-T cells in the clinic is the autoimmune reaction against self-tissues produced by "off-target" toxicity [32]. Most selectively targeted tumor cell antigens have overlapping expression on normal tissues, and this can cause potential off-target effects by generating a too potent and too effective immune response that results in a cytokine storm inside the patient's body and ultimately lead to a fatal outcome [33]. This off-target effect is specifically a major concern with the second and third-generation CAR-T cells. Scientists are currently trying to develop means to overcome this hurdle by generating CARs displaying a lower affinity to their target antigen, and several research groups are currently working on generating CARs that show lower affinity binding to their target tumor cell antigens [34, 35].

7 Concluding Remarks

With this chapter, an attempt was made to highlight the essential features of CAR-T cell development, which contribute toward successful tumor immunotherapy, namely selecting the specific tumor cell antigen to target, engineering the signaling domain for maximum efficiency, ability to overcome immune evasion by tumor cells, ability to localize to disease sites, ability to persist and expand in vivo, and finally the ability to retain tumor cell lytic function. While clinical trials with the first-generation CARs showed modest antitumor efficacy, these initial trials paved the way for further improvements of CAR function and CAR design. This culminated in the engineering of second- and third-generation CARs capable of delivering greater strength and quality activation signals that directly resulted in enhanced antitumor properties. So far, clinical reports show tremendous promise and potential for tumor management but also highlight challenges to overcome in order to advance toward an effective and safe cancer therapy.

References

1. Eshhar, Z.: Cancer Immunol. Immunother.: CII **45**, 131–136 (1997)
2. (a) Hicklin, D.J., Marincola, F.M., Ferrone, S.: HLA class I antigen downregulation in human cancers: T-cell immunotherapy revives an old story. Mol. Med. Today **5**, 178–186 (1999); (b) Rivoltini, L., Carrabba, M., Huber, V., Castelli, C., Novellino, L., Dalerba, P., Mortarini, R., Arancia, G., Anichini, A., Fais, S., Parmiani, G.: Immunity to cancer: attack and escape in T lymphocyte–tumor cell interaction. Immunol. Rev. **188**, 97–113 (2002)
3. (a) Sadelain, M., Riviere, I., Brentjens, R.: Targeting tumours with genetically enhanced T lymphocytes. Nat. Rev. Cancer **3**, 35–45 (2003); (b) Bridgeman, J.S., Hawkins, R.E., Hombach, A.A., Abken, H., Gilham, D.E.: Building better chimeric antigen receptors for adoptive T cell therapy. Curr. Gene Ther. **10**, 77–90 (2010)
4. (a) Niederman, T.M., Ghogawala, Z., Carter, B.S., Tompkins, H.S., Russell, M.M., Mulligan, R.C.: Antitumor activity of cytotoxic T lymphocytes engineered to target vascular endothelial growth factor receptors. Proc. Natl. Acad. Sci. U.S.A. **99**, 7009–7014 (2002); (b) Pameijer, C.R., Navanjo, A., Meechoovet, B., Wagner, J.R., Aguilar, B., Wright, C.L., Chang, W.C., Brown, C.E., Jensen, M.C.: Conversion of a tumor-binding peptide identified by phage display to a functional chimeric T cell antigen receptor. Cancer Gene Ther. **14**, 91–97 (2007); (c) Stastny, M.J., Brown, C.E., Ruel, C., Jensen, M.C.: Medulloblastomas expressing IL13Rα2 are targets for IL13-zetakine + Cytolytic T cells. J. Pediatr. Hematol./Oncol. **29**, 669–677 (2007); (d) Kahlon, K.S., Brown, C., Cooper, L.J., Raubitschek, A., Forman, S.J., Jensen, M.C.: Specific recognition and killing of glioblastoma multiforme by interleukin 13-zetakine redirected cytolytic T cells. Cancer Res. **64**, 9160–9166 (2004); (e) Muniappan, A., Banapour, B., Lebkowski, J., Talib, S.: Ligand-mediated cytolysis of tumor cells: use of heregulin-ζ chimeras to redirect cytotoxic T lymphocytes. Cancer Gene Ther. **7**, 128–134 (2000); (f) Zhang, T., Lemoi, B.A., Sentman, C.L.: Chimeric NK-receptor–bearing T cells mediate antitumor immunotherapy. Blood **106**, 1544–1551 (2005); (g) Zhang T., Barber, A., Sentman, C.L.: Chimeric NK-receptor–bearing T cells mediate antitumor immunotherapy. Cancer Res. **67**, 11029–11036 (2007); (h) Barber, A., Zhang, T., Sentman, C.L.: Immunotherapy with Chimeric NKG2D receptors leads to long-term tumor-free survival and development of host antitumor immunity in murine ovarian cancer. J. Immunol. **180**, 72–78 (2008)
5. (a) Moritz, D., Groner, B.: A spacer region between the single chain antibody-and the CD3 zeta-chain domain of chimeric T cell receptor components is required for efficient ligand binding and signaling activity. Gene Ther. **2**, 539–546 (1995); (b) Wilkie, S., Picco, G., Foster, J., Davies, D.M., Julien, S., Cooper, L., Arif, S., Mather, S.J., Taylor-Papadimitriou, J., Burchell, J.M., Maher, J.: Retargeting of human T cells to tumor-associated MUC1: the evolution of a chimeric antigen receptor. J. Immunol. **180**, 4901–4909 (2008); (c) Hombach, A., Hombach, A.A., Abken, H.: Adoptive immunotherapy with genetically engineered T cells: modification of the IgG1 Fc 'spacer' domain in the extracellular moiety of chimeric antigen receptors avoids 'off-target' activation and unintended initiation of an innate immune response. Gene Ther. **17**, 1206–1213 (2010); (d) Kradin, R., Kurnick, J., Gifford, J., Pinto, C., Preffer, F., Lazarus, D.: Adoptive immunotherapy with interleukin-2 (IL-2) results in diminished IL-2 production by stimulated peripheral blood lymphocytes. J. Clin. Immunol. **9**, 378–385 (1989)
6. (a) Milone, M.C., Fish, J.D., Carpenito, C., Carroll, R.G., Binder, G.K., Teachey, D., Samanta, M., Lakhal, M., Gloss, B., Danet-Desnoyers, G., Campana, D., Riley, J.L., Grupp, S.A., June, C.H.: Mol. Ther.: Chimeric receptors containing CD137 signal transduction domains mediate enhanced survival of T cells and increased antileukemic efficacy in vivo. J. Am. Soc. Gene Ther. **17**, 1453–1464 (2009); (b) Hombach, A.A., Heiders, J., Foppe, M., Chmielewski, M., Abken, H.: OX40 costimulation by a chimeric antigen receptor abrogates CD28 and IL-2 induced IL-10 secretion by redirected $CD4^+$ T cells. Oncoimmunology **1**, 458–466 (2012); (c) Song, D.G., Ye, Q., Carpenito, C., Poussin, M., Wang, L.P., Ji, C., Figini, M., June, C.H., Coukos, G., Powell, D.J., Jr.: In vivo persistence, tumor localization, and antitumor activity of CAR-engineered T cells is enhanced by costimulatory signaling through CD137 (4-1BB). Cancer

Res. **71**, 4617–4627 (2011); (d) Tammana, S., Huang, X., Wong, M., Milone, M.C., Ma, L., Levine, B.L., June, C.H., Wagner, J.E., Blazar, B.R., Zhou, X.: 4-1BB and CD28 signaling plays a synergistic role in redirecting umbilical cord blood T cells against B-cell malignancies. Human Gene Ther. **21**, 75–86 (2010); (e) Pule, M.A., Straathof, K.C., Dotti, G., Heslop, H.E., Rooney, C.M., Brenner, M.K.: Mol. Ther.: A chimeric T cell antigen receptor that augments cytokine release and supports clonal expansion of primary human T cells. J. Am. Soc. Gene Ther. **12**, 933–941 (2005)

7. Irving, B.A., Weiss, A.: The cytoplasmic domain of the T cell receptor ζ chain is sufficient to couple to receptor-associated signal transduction pathways. Cell **64**, 891–901 (1991)
8. (a) Letourneur, F., Klausner, R.D.: T-cell and basophil activation through the cytoplasmic tail of T-cell-receptor zeta family proteins. Proc. Natl. Acad. Sci. U. S. A. **88**, 8905–8909 (1991); (b) Romeo, C., Amiot, M., Seed, B.: Sequence requirements for induction of cytolysis by the T cell antigenFc receptor ζ chain. Cell **68**, 889–897 (1992)
9. (a) Zhong, X.S., Matsushita, M., Plotkin, J., Riviere, I., Sadelain, M.: Mol. Ther.: Chimeric antigen receptors combining 4-1BB and CD28 signaling domains augment PI3kinase/AKT/Bcl-XL activation and CD8 + T cell–mediated tumor eradication. J. Am. Soc. Gene Ther. **18**, 413–420 (2010); (b) Wang, L.X., Westwood, J.A., Moeller, M., Duong, C.P., Wei, W.Z., Malaterre, J., Trapani, J.A., Neeson, P., Smyth, M.J., Kershaw, M.H., Darcy, P.K.: Cancer Res. **70**, 9591–9598 (2010); (c) Carpenito, C., Milone, M.C., Hassan, R., Simonet, J.C., Lakhal, M., Suhoski, M.M., Varela-Rohena, A., Haines, K.M., Heitjan, D.F., Albelda, S.M., Carroll, R.G., Riley, J.L., Pastan, I., June, C.H.: Control of large, established tumor xenografts with genetically retargeted human T cells containing CD28 and CD137 domains. Proc. Natl. Acad. Sci. U. S. A. **106**, 3360–3365 (2009)
10. (a) Imai, C., Mihara, K., Andreansky, M., Nicholson, I.C., Pui, C.H., Geiger, T.L., Campana, D.: Chimeric receptors with 4-1BB signaling capacity provoke potent cytotoxicity against acute lymphoblastic leukemia. Leukemia **18**, 676–684 (2004); (b) Maher, J., Brentjens, R.J., Gunset, G., Riviere, I., Sadelain, M.: Human T-lymphocyte cytotoxicity and proliferation directed by a single chimeric TCRζ/CD28 receptor. Nat. Biotechnol. **20**, 70–75 (2002); (c) Kowolik, C.M., Topp, M.S., Gonzalez, S., Pfeiffer, T., Olivares, S., Gonzalez, N., Smith, D.D., Forman, S.J., Jensen, M.C., Cooper, L.J.: CD28 costimulation provided through a CD19-specific chimeric antigen receptor enhances in vivo persistence and antitumor efficacy of adoptively transferred T cells. Cancer Res. **66**, 10995–11004 (2006); (d) Finney, H.M., Akbar, A.N., Lawson, A.D.: Activation of resting human primary T cells with chimeric receptors: costimulation from CD28, inducible costimulator, CD134, and CD137 in series with signals from the TCRζ chain. J. Immunol. **172**, 104–113 (2004); (e) Song, D.G., Ye, Q., Poussin, M., Harms, G.M., Figini, M., Powell, D.J., Jr.: CD27 costimulation augments the survival and antitumor activity of redirected human T cells in vivo. Blood **119**, 696–706 (2012); (f) Altvater, B., Landmeier, S., Pscherer, S., Temme, J., Juergens, H., Pule, M., Rossig, C.: Cancer Immunol. Immunother.: CII **58**, 1991–2001 (2009)
11. (a) Pinthus, J.H., Waks, T., Kaufman-Francis, K., Schindler, D.G., Harmelin, A., Kanety, H., Ramon, J., Eshhar, Z.: Immuno-gene therapy of established prostate tumors using chimeric receptor-redirected human lymphocytes. Cancer Res. **63**, 2470–2476 (2003); (b) Teng, M.W., Kershaw, M.H., Moeller, M., Smyth, M.J., Darcy, P.K.: Immunotherapy of cancer using systemically delivered gene-modified human T lymphocytes. Human Gene Ther. **15**, 699–708 (2004); (c) Gong, M.C., Latouche, J.B., Krause, A., Heston, W.D., Bander, N.H., Sadelain, M.: Cancer patient T cells genetically targeted to prostate-specific membrane antigen specifically lyse prostate cancer cells and release cytokines in response to prostate-specific membrane antigen. Neoplasia **1**, 123–127 (1999); (d) Weijtens, M.E., Willemsen, R.A., van Krimpen, B.A., Bolhuis, R.L.: Chimeric scFv/γ receptor-mediated T-cell lysis of tumor cells is coregulated by adhesion and accessory molecules. Int. J. Cancer **77**, 181–187 (1998); (e) Lamers, C.H., Sleijfer, S., Vulto, A.G., Kruit, W.H., Kliffen, M., Debets, R., Gratama, J.W., Stoter, G., Oosterwijk, E.: Treatment of metastatic renal cell carcinoma with autologous T-lymphocytes genetically retargeted against carbonic anhydrase IX: first clinical experience. J. Clin. Oncol.: Official J. Am. Soc. Clin. Oncol. **24**, e20–22 (2006); (f) Mezzanzanica, D., Canevari, S., Mazzoni, A.,

Figini, M., Colnaghi, M.I., Waks, T., Schindler, D.G., Eshhar, Z.: Transfer of chimeric receptor gene made of variable regions of tumor-specific antibody confers anticarbohydrate specificity on T cells. Cancer Gene Ther. **5**, 401–407 (1998); (g) Westwood, J.A., Smyth, M.J., Teng, M.W., Moeller, M., Trapani, J.A., Scott, A.M., Smyth, F.E., Cartwright, G.A., Power, B.E., Honemann, D., Prince, H.M., Darcy, P.K., Kershaw, M.H.: Adoptive transfer of T cells modified with a humanized chimeric receptor gene inhibits growth of Lewis-Y-expressing tumors in mice. Proc. Natl. Acad. Sci. U. S. Am. **102**, 19051–19056 (2005); (h) Darcy, P.K., Kershaw, M.H., Trapani, J.A., Smyth, M.J.: Expression in cytotoxic T lymphocytes of a single-chain anti-carcinoembryonic antigen antibody. Redirected fas ligand-mediated lysis of colon carcinoma. Eur. J. Immunol. **28**, 1663–1672 (1998); (i) Haynes, N.M., Trapani, J.A., Teng, M.W., Jackson, J.T., Cerruti, L., Jane, S.M., Kershaw, M.H., Smyth, M.J., Darcy, P.K.: Rejection of syngeneic colon carcinoma by CTLs expressing single-chain antibody receptors codelivering CD28 costimulation. J. Immunol. **169**, 5780–5786 (2002); (j) Hwu, P., Shafer, G.E., Treisman, J., Schindler, D.G., Gross, G., Cowherd, R., Rosenberg, S.A., Eshhar, Z.: Lysis of ovarian cancer cells by human lymphocytes redirected with a chimeric gene composed of an antibody variable region and the Fc receptor gamma chain. J. Exp. Med. **178**, 361–366 (1993); (k) Chekmasova, A.A., Rao, T.D., Nikhamin, Y., Park, K.J., Levine, D.A., Spriggs, D.R., Brentjens, R.J.: Clin. Cancer Res.: Official J. Am. Assoc. Cancer Res. **16**, 3594–3606 (2010); (l) Rossig, C., Bollard, C.M., Nuchtern, J.G., Rooney, C.M., Brenner, M.K.: Epstein-Barr virus–specific human T lymphocytes expressing antitumor chimeric T-cell receptors: potential for improved immunotherapy. Blood **99**, 2009–2016 (2002)

12. Petrausch, U., Jensen, S.M., Twitty, C., Poehlein, C.H., Haley, D.P., Walker, E.B., Fox, B.A.: Disruption of TGF-β signaling prevents the generation of tumor-sensitized regulatory T cells and facilitates therapeutic antitumor immunity. J. Immunol. **183**, 3682–3689 (2009)
13. (a) Foster, A.E., Dotti, G., Lu, A., Khalil, M., Brenner, M.K., Heslop, H.E., Rooney, C.M., Bollard, C.M.: Antitumor activity of EBV-specific T lymphocytes transduced with a dominant negative TGF-β receptor. J. Immunother. **31**, 500–505 (2008); (b) Bollard, C.M., Rossig, C., Calonge, M.J., Huls, M.H., Wagner, H.J., Massague, J., Brenner, M.K., Heslop, H.E., Rooney, C.M.: Adapting a transforming growth factor β–related tumor protection strategy to enhance antitumor immunity. Blood **99**, 3179–3187 (2002)
14. Eshhar, Z., Waks, T., Gross, G., Schindler, D.G.: Specific activation and targeting of cytotoxic lymphocytes through chimeric single chains consisting of antibody-binding domains and the gamma or zeta subunits of the immunoglobulin and T-cell receptors. Proc. Natl. Acad. Sci. U. S. A. **90**, 720–724 (1993)
15. (a) Heuser, C., Hombach, A., Losch, C., Manista, K., Abken, H.: T-cell activation by recombinant immunoreceptors: impact of the intracellular signalling domain on the stability of receptor expression and antigen-specific activation of grafted T cells. Gene Ther. **10**, 1408–1419 (2003); (b) Haynes, N.M., Snook, M.B., Trapani, J.A., Cerruti, L., Jane, S.M., Smyth, M.J., Darcy, P.K.: Redirecting mouse CTL against colon carcinoma: superior signaling efficacy of single-chain variable domain chimeras containing TCR-ζ vs FcεRI-γ. J. Immunol. **166**, 182–187 (2001)
16. (a) Harding, F.A., McArthur, J.G., Gross, J.A., Raulet, D.H., Allison, J.P.: CD28-mediated signalling co-stimulates murine T cells and prevents induction of anergy in T-cell clones. Nature **356**, 607–609 (1992); (b) Lenschow, D.J., Walunas, T.L., Bluestone, J.A.: CD28/B7 system of T cell costimulation. Ann. Rev. Immunol. **14**, 233–258 (1996)
17. (a) Smith-Garvin, J.E., Koretzky, G.A., Jordan, M.S.: T cell activation. Ann. Rev. Immunol. **27**, 591–619 (2009); (b) Chambers, C.A., Allison, J.P.: Costimulatory regulation of T cell function. Curr. Opin. Cell Biol. **11**, 203–210 (1999)
18. (a) Bretscher, P., Cohn, M.: A theory of self-nonself discrimination: paralysis and induction involve the recognition of one and two determinants on an antigen, respectively. Science **169**, 1042–1049 (1970); (b) Bretscher, P.A.: A two-step, two-signal model for the primary activation of precursor helper T cells. Proc. Natl. Acad. Sci. U. S. A. **96**, 185–190 (1999)
19. (a) Beecham, E.J., Ma, Q., Ripley, R., Junghans, R.P.: Coupling CD28 co-stimulation to immunoglobulin T-cell receptor molecules: the dynamics of T-cell proliferation and death. J. Immunother. **23**, 631–642 (2000); (b) Hombach, A., Sent, D., Schneider, C., Heuser, C.,

Koch, D., Pohl, C., Seliger, B., Abken, H.: T-cell activation by recombinant receptors: CD28 costimulation is required for interleukin 2 secretion and receptor-mediated T-cell proliferation but does not affect receptor-mediated target cell lysis. Cancer Res. **61**, 1976–1982 (2001); (c) Hombach, A., Wieczarkowiecz, A., Marquardt, T., Heuser, C., Usai, L., Pohl, C., Seliger, B., Abken, H.: Tumor-specific T cell activation by recombinant immunoreceptors: CD3ζ signaling and CD28 costimulation are simultaneously required for efficient IL-2 secretion and can be integrated into one combined CD28/CD3ζ signaling receptor molecule. J. Immunol. **167**, 6123–6131 (2001); (d) Vera, J., Savoldo, B., Vigouroux, S., Biagi, E., Pule, M., Rossig, C., Wu, J., Heslop, H.E., Rooney, C.M., Brenner, M.K., Dotti, G.: T lymphocytes redirected against the κ light chain of human immunoglobulin efficiently kill mature B lymphocyte-derived malignant cells. Blood **108**, 3890–3897 (2006)

20. (a) Brentjens, R.J., Latouche, J.B., Santos, E., Marti, F., Gong, M.C., Lyddane, C., King, P.D., Larson, S., Weiss, M., Riviere, I., Sadelain, M.: Nat. Med. **9**, 279–286 (2003); (b) Gilham, D.E., O'Neil, A., Hughes, C., Guest, R.D., Kirillova, N., Lehane, M., Hawkins, R.E.: J. Immunother. **25**, 139–151 (2002); (c) Stancovski, I., Schindler, D.G., Waks, T., Yarden, Y., Sela, M., Eshhar, Z.: J. Immunol. **151**, 6577–6582 (1993)
21. Santos, E.B., Yeh, R., Lee, J., Nikhamin, Y., Punzalan, B., La Perle, K., Larson, S.M., Sadelain, M., Brentjens, R.J.: Sensitive in vivo imaging of T cells using a membrane-bound Gaussia princeps luciferase. Nat. Med. **15**, 338–344 (2009)
22. Savoldo, B., Ramos, C.A., Liu, E., Mims, M.P., Keating, M.J., Carrum, G., Kamble, R.T., Bollard, C.M., Gee, A.P., Mei, Z., Liu, H., Grilley, B., Rooney, C.M., Heslop, H.E., Brenner, M.K., Dotti, G.: CD28 costimulation improves expansion and persistence of chimeric antigen receptor–modified T cells in lymphoma patients. J. Clin. Invest. **121**, 1822–1826 (2011)
23. (a) Brentjens, R.J., Riviere, I., Park, J.H., Davila, M.L., Wang, X., Stefanski, J., Taylor, C., Yeh, R., Bartido, S., Borquez-Ojeda, O., Olszewska, M., Bernal, Y., Pegram, H., Przybylowski, M., Hollyman, D., Usachenko, Y., Pirraglia, D., Hosey, J., Santos, E., Halton, E., Maslak, P., Scheinberg, D., Jurcic, J., Heaney, M., Heller, G., Frattini, M., Sadelain, M.: Safety and persistence of adoptively transferred autologous CD19-targeted T cells in patients with relapsed or chemotherapy refractory B-cell leukemias. Blood **118**, 4817–4828 (2011); (b) Kalos, M., Levine, B.L., Porter, D.L., Katz, S., Grupp, S.A., Bagg, A., June, C.H.: T cells with chimeric antigen receptors have potent antitumor effects and can establish memory in patients with advanced leukemia. Sci. Transl. Med. **3**, 95ra73 (2011)
24. (a) Kershaw, M.H., Wang, G., Westwood, J.A., Pachynski, R.K., Tiffany, H.L., Marincola, F.M., Wang, E., Young, H.A., Murphy, P.M., Hwu, P.: Redirecting migration of T cells to chemokine secreted from tumors by genetic modification with CXCR2. Human Gene Ther. **13**, 1971–1980 (2002); (b) Di Stasi, A., De Angelis, B., Rooney, C.M., Zhang, L., Mahendravada, A., Foster, A.E., Heslop, H.E., Brenner, M.K., Dotti, G., Savoldo, B.: T lymphocytes coexpressing CCR4 and a chimeric antigen receptor targeting CD30 have improved homing and antitumor activity in a Hodgkin tumor model. Blood **113**, 6392–6402 (2009)
25. Kochenderfer, J.N., Wilson, W.H., Janik, J.E., Dudley, M.E., Stetler-Stevenson, M., Feldman, S.A., Maric, I., Raffeld, M., Nathan, D.A., Lanier, B.J., Morgan, R.A., Rosenberg, S.A.: Eradication of B-lineage cells and regression of lymphoma in a patient treated with autologous T cells genetically engineered to recognize CD19. Blood **116**, 4099–4102 (2010)
26. (a) Schreiber, R.D., Old, L.J., Smyth, M.J.: Cancer immunoediting: integrating immunity's roles in cancer suppression and promotion. Science **331**, 1565–1570 (2011); (b) Rabinovich, G.A., Gabrilovich, D., Sotomayor, E.M.: Immunosuppressive strategies that are mediated by tumor cells. Ann. Rev. Immunol. **25**, 267–296 (2007)
27. (a) Park, J.R., Digiusto, D.L., Slovak, M., Wright, C., Naranjo, A., Wagner, J., Meechoovet, H.B., Bautista, C., Chang, W.C., Ostberg, J.R., Jensen, M.C.: Mol. Ther.: Adoptive transfer of chimeric antigen receptor re-directed cytolytic T lymphocyte clones in patients with neuroblastoma. J. Am. Soc. Gene Ther. **15**, 825–833 (2007); (b) Till, B.G., Jensen, M.C., Wang, J., Chen, E.Y., Wood, B.L., Greisman, H.A., Qian, X., James, S.E., Raubitschek, A., Forman, S.J., Gopal, A.K., Pagel, J.M., Lindgren, C.G., Greenberg, P.D., Riddell, S.R., Press, O.W.: Adoptive

immunotherapy for indolent non-Hodgkin lymphoma and mantle cell lymphoma using genetically modified autologous CD20-specific T cells. Blood **112**, 2261–2271 (2008); (c) Kershaw, M.H., Westwood, J.A., Parker, L.L., Wang, G., Eshhar, Z., Mavroukakis, S.A., White, D.E., Wunderlich, J.R., Canevari, S., Rogers-Freezer, L., Chen, C.C., Yang, J.C., Rosenberg, S.A., Hwu, P.: A phase I study on adoptive immunotherapy using gene-modified T cells for ovarian cancer. Clin. Cancer Res.: Official J. Am. Assoc. Cancer Res. **12**, 6106–6115 (2006)
28. (a) Louis, C.U., Savoldo, B., Dotti, G., Pule, M., Yvon, E., Myers, G.D., Rossig, C., Russell, H.V., Diouf, O., Liu, E., Liu, H., Wu, M.F., Gee, A.P., Mei, Z., Rooney, C.M., Heslop, H.E., Brenner, M.K.: Anti-tumor activity and long-term fate of chimeric antigen receptor positive T-cells in patients with neuroblastoma. Blood **118**, 6050–6056 (2011); (b) Pule, M.A., Savoldo, B., Myers, G.D., Rossig, C., Russell, H.V., Dotti, G., Huls, M.H., Liu, E., Gee, A.P., Mei, Z., Yvon, E., Weiss, H.L., Liu, H., Rooney, C.M., Heslop, H.E., Brenner, M.K.: Virus-specific T cells engineered to coexpress tumor-specific receptors: persistence and antitumor activity in individuals with neuroblastoma. Nat. Med. **14**, 1264–1270 (2008); (c) Lamers, C.H., Willemsen, R., van Elzakker, P., van Steenbergen-Langeveld, S., Broertjes, M., Oosterwijk-Wakka, J., Oosterwijk, E., Sleijfer, S., Debets, R., Gratama, J.W.: Blood **117**, 72–82 (2011); (d) Jensen, M.C., Popplewell, L., Cooper, L.J., DiGiusto, D., Kalos, M., Ostberg, J.R., Forman, S.J.: Antitransgene rejection responses contribute to attenuated persistence of adoptively transferred CD20/CD19-specific chimeric antigen receptor redirected T cells in humans. Biol. Blood Marrow Transplant.: J. Am. Soc. Blood Marrow Transplant. **16**, 1245–1256 (2010)
29. Morgan, R.A., Yang, J.C., Kitano, M., Dudley, M.E., Laurencot, C.M., Rosenberg, S.A.: Case report of a serious adverse event following the administration of T cells transduced with a chimeric antigen receptor recognizing ERBB2. Molecular therapy: the journal of the American Society of Gene Therapy **18**, 843–851 (2010)
30. (a) Hollyman, D., Stefanski, J., Przybylowski, M., Bartido, S., Borquez-Ojeda, O., Taylor, C., Yeh, R., Capacio, V., Olszewska, M., Hosey, J., Sadelain, M., Brentjens, R.J., Riviere, I.: Manufacturing validation of biologically functional T cells targeted to CD19 antigen for autologous adoptive cell therapy. J. Immunother **32**, 169–180 (2009); (b) June, C.H.: Adoptive T cell therapy for cancer in the clinic. J. Clin. Investig. **117**, 1466–1476 (2007)
31. (a) Seliger, B.: Different regulation of MHC class I antigen processing components in human tumors. J. Immunotoxicol. **5,** 361–367 (2008); (b) Kerkar, S.P., Muranski, P., Kaiser, A., Boni, A., Sanchez-Perez, L., Yu, Z., Palmer, D.C., Reger, R.N., Borman, Z.A., Zhang, L., Morgan, R.A., Gattinoni, L., Rosenberg, S.A., Trinchieri, G., Restifo, N.P.: Tumor-specific CD8+T cells expressing interleukin-12 eradicate established cancers in lymphodepleted hosts. Cancer Res. **70,** 6725–6734 **(2010)**
32. (a) Casucci, M., Hawkins, R.E., Dotti, G., Bondanza, A.: Overcoming the toxicity hurdles of genetically targeted T cells. Cancer Immunol. Immunother. CII **64**, 123–130 **(2015)**; (b) Grupp, S.A., Kalos, M., Barrett, D., Aplenc, R., Porter, D.L., Rheingold, S.R., Teachey, D.T., Chew, A., Hauck, B., Wright, J.F., Milone, M.C., Levine B.L., June, C.H.: Chimeric antigen receptor–modified T cells for acute lymphoid leukemia. New England J. Med. **368**, 1509–1518 (2013); (c) Linette, G.P., Stadtmauer, E.A., Maus, M.V., Rapoport, A.P., Levine, B.L., Emery, L., Litzky, L., Bagg, A., Carreno, B.M., Cimino, P.J., Binder-Scholl, G.K., Smethurst, D.P., Gerry, A.B., Pumphrey, N.J., Bennett, A.D., Brewer, J.E., Dukes, J., Harper, J., Tayton-Martin, H.K., Jakobsen, B.K., Hassan, N.J., Kalos, M., June, C.H.: Cardiovascular toxicity and titin cross-reactivity of affinity enhanced T cells in myeloma and melanoma. Blood **122,** 863–871 (2013); (d) Cameron, B.J., Gerry, A.B., Dukes, J., Harper, J.V., Kannan, V., Bianchi, F.C., Grand, F., Brewer, J.E., Gupta, M., Plesa, G., Bossi, G., Vuidepot, A., Powlesland, A.S., Legg, A., Adams, K.J., Bennett, A.D., Pumphrey, N.J., Williams, D.D., Binder-Scholl, G., Kulikovskaya, I., Levine, B.L., Riley, J.L., Varela-Rohena, A., Stadtmauer, E.A., Rapoport, A.P., Linette, G.P., June, C.H., Hassan, N.J., Kalos, M., Jakobsen, B.K.: Identification of a Titin-derived HLA-A1–presented peptide as a cross-reactive target for engineered MAGE A3–directed T cells. Science Transl. Med. **5**, 197ra103 (2013)
33. Lee, D.W., Gardner, R., Porter, D.L., Louis, C.U., Ahmed, N., Jensen, M., Grupp, S.A., Mackall, C.L.: How I treat: current concepts in the diagnosis and management of cytokine release

syndrome. Blood **124**, 188–195 (2014)

34. Ramos, C.A., Asgari, Z., Liu, E., Yvon, E., Heslop, H.E., Rooney, C.M., Brenner, M.K., Dotti, G.: An inducible caspase 9 suicide gene to improve the safety of mesenchymal stromal cell therapies. Stem Cells **28**, 1107–1115 (2010)
35. Ciceri, F., Bonini, C., Stanghellini, M.T., Bondanza, A., Traversari, C., Salomoni, M., Turchetto, L., Colombi, S., Bernardi, M., Peccatori, J., Pescarollo, A., Servida, P., Magnani, Z., Perna, S.K., Valtolina, V., Crippa, F., Callegaro, L., Spoldi, E., Crocchiolo, R., Fleischhauer, K., Ponzoni, M., Vago, L., Rossini, S., Santoro, A., Todisco, E., Apperley, J., Olavarria, E., Slavin, S., Weissinger, E.M., Ganser, A., Stadler, M., Yannaki, E., Fassas, A., Anagnostopoulos, A., Bregni, M., Stampino, C.G., Bruzzi, P., Bordignon, C.: Tk007 phase Ii study final analysis. Genetically modified donor lymphocytes abate transplant related mortality and increase leukaemia-free survival providing early immune reconstitution after haploidentical haematopoietic stem cell transplantation. Lancet Oncol **10**, 489–500 (2009)

Part IV
Greener Environment

Performance Evaluation of Solar Energy-Based Distillation System for Groundwater Purification: A Green Concept for Rural Development of Indian Villages

Shivakshi Jasrotia

Abstract Globally, increasing water pollution and lack of safe drinking water has raised an alarming situation. There are breaching water treatment systems which are increasing the rate of mortality by severely affecting the health and severe health effects due to lack of potable water. This guides us to develop and initiate sustainable solutions for both urban and rural communities. A common solution for many water-related key issues like an assurance to its availability, its reliability, and purity can be dealt with the development of low-tech systems which reflect and promote long-term sustainable solutions. The following chapter discusses a process that harnesses the solar energy for the removal of arsenic contamination in water, followed by phytoremediation to treat the generated waste to meet the disposal requirements. The chapter also throws light on alternative solutions relying on solar energy for rural development, with economic evaluation to address the vulnerability of residents in context to the changing environment, climate change, and groundwater pollution, etc. The chapter opens new roads of environmental sciences taking the lead for agricultural biotechnological problems of groundwater contamination.

Keywords Groundwater pollution · Solar energy · Public health · Remediation Rural development · Climate change and vulnerability

1 Introduction

Sustainability means the use of natural resources with principle of—equity, efficiency, optimization, and decentralized planning. Resources include both natural

S. Jasrotia (✉)
Earth Science and Climate Change Division, Centre for Global Environmental Research, The Energy and Resources Institute (TERI), IHC Complex, Lodhi Road, New Delhi 110003, India
e-mail: shivakshi_jasrotia@yahoo.com; s.jasrotia@teri.res.in

M. Khoobchandani and A. Saxena (eds.), *Biotechnology Products in Everyday Life*, EcoProduction, https://doi.org/10.1007/978-3-319-92399-4_8

endowments and man-made. Water is one of the most fundamental and versatile resources among all natural resources. It is both unitary and finite with few substitutes or no substitute for many applications. Indiscriminate use and toxic discharges have polluted groundwater and made it unfit for human consumption. "Water and health" is the theme that poses major challenge to sustainability of man on earth. Nearly, 1 billion people in developing countries, particularly the rural poor, lack access to potable water and 2.6 billion has to contend with inadequate sanitation facilities [1]. It is now globally accepted that water- and sanitation-related diseases are the single most cause of mortality and morbidity. The two inseparable aspects of sustainable water management are water quantity and its quality. In recent times, other than microbial diseases, chemical characteristics of water, such as arsenic metal, fluorides, and nitrates are increasingly being recognized. Among the present contaminants in drinking water, the risk of dissolved arsenic in its various forms is life-threatening. The factors responsible for this include industrial discharge, climate change, and the large variability in rainfall and hence more dependence on groundwater sources with the large-scale abstraction of groundwater (at times more than that of recharge potential). The risks associated with consumption of arsenic and nitrates, etc., and affected drinking water are both toxicological as well as environmental. Coincidentally, most of the arsenic-affected areas are rural and the economically poor population is the most vulnerable.

The menace of arsenic has been reported very high in groundwater especially in countries like India, Bangladesh, Pakistan, and Nepal. The condition is such that these countries have the elevated levels more than 10 times in comparison to the drinking water standards as mentioned by WHO of 0.01 mg/L [2]. Nearly, 100 million people are at a health risk due to the elevated arsenic metal presence in groundwater [3]. From the 33–77 million affected patients, approximately 70 million people are in India itself who are vulnerable and residing in arsenic affected areas, respectively [4, 5]. Numerous methods for its treatment have been investigated time and again like coagulation; filtration and ion-exchange for arsenic removal; methods like these have been developed and tested on the field. But, ironically, it has also been mentioned in the literature that these methods either need electricity to run or require monitoring of certain performance parameters on a regular basis. They are also found to generate hazardous waste that limits the sustained performance of these technologies in rural field application. Now in context to a rural area that has no access to skilled manpower and electricity, these treatment alternatives will be defunct and futile with time. The following three major barriers have been highlighted which make safe drinking water supply and sanitation programs in rural areas unsuccessful over the time. Firstly, due to focus on other major priorities, the rural infrastructure projects in low- and middle-income countries are found to be financially unattractive. Secondly, an uninterrupted electrical form of energy for 100% operation of treatment units is needed for most of the water treatment process (energy intensive). Thirdly, the reliability of treated water gets questionable in such systems based on the adsorption method as most of the natural systems suffer from the lack of information about the completion of media or its replacement. Therefore, sustainable management of water in rural areas which is consistent and reliable requires an innovative approach that takes into account ground realities before operation. In light of this, the present study aims to provide a renewable energy-based cost-effective and sustainable solution for drinking water and sanitation in rural areas. One such approach is exploring and harnessing solar energy for this purpose. The solar energy on earth is most efficiently tapped in two forms which are—solar energy as heat (thermal capturing) and solar capturing through plant biomass-based remediation and biomass production (photon or radiation capturing).

The objective of this research is to not only to address this issue through the use of solar energy for water purification in the context of rural water supply and removal of metal like arsenic, but to also evaluate the sustainability equation in terms of application. This research study outlines and investigates a concept for drinking water supply in rural household, which is provided through the process of (1) capturing solar heat using solar still technology, (2) the wastewater generated from solar still (brine) is disposed after post-treatment options like plant biomass remediation.

2 Solar Distillation and Water Purification: Historical Perspective and Types

Water and energy are two indivisible items that have governed our lives and promoted civilization [6]. Desalination is the oldest technology and has been used for water purification globally especially salt water to drinking water. Between the several possible options for desalination, renewable energy technology is the most promising technique for water purification in terms of economic stability and technological feasibility [7]. The various technologies that have been invented and employed for desalination purpose include vapor compression distillation, reverse osmosis, and electrodialysis where electric energy is the source energy. Considering the rising energy crises, renewable energy-driven desalination technologies such as distillation desalination using solar energy becomes a widely acceptable solution.

Distillation has been considered as a way of making salt water drinkable from sources in remote locations [8]. The history of using solar energy to treat water dates back to the era of the Renaissance, when desalination was the natural method used to treat brackish water [6]. It was Arab alchemists in the sixteenth century, who were the first known to have used the solar distillation system for treating water [8]. Wheeler and Evans received the first American patent in 1870 for solar distillation [6, 9]. Later on, after two years, in 1872, the first large solar still was built in Las Salinas, Chile, by Carlos Wilson, which had 64 basins designed from wood and timber framework with sloping glass covers [6, 8]. From 1950 onwards, developments have been made in solar still for developing larger plants to treat water globally. In 1978, Bhavnagar got the first largest solar distillation plant with 90 still units, installed by Central Salt and Marine Chemical Research Institute to supply drinking water to non-electrified Awania village of the system are (1) raw water, (2) enough sunlight, and (3) black

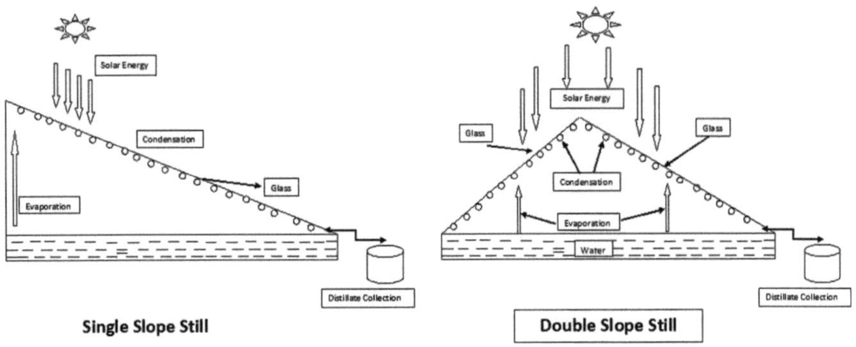

Fig. 1 Single and double slope still

basin body. Unlike a community water supply system, it does not require any other infrastructure like water pipelines, heavy maintenance, and Chhachi lighthouse [8].

Solar stills are sealed units with glass cover, which evaporate brine/brackish water for household or large community supplies [9]. As these stills have the capacity to produce limited quantities of drinking water for a household and require no grid electricity or skilled labor or heavy technology, they can be a viable solution in rural areas where above-mentioned limitations are not a constraint [9, 10].

Numerous types of solar still like single slope, double slope multi-basin (Fig. 1) have been reviewed by various researchers. [11, 12]. Among the existing solar still types like conventional type, single and multiple basin, wick type, multiple effect, basin-type stills have been used for the supply of large and small quantities of water [11]. In this study, single slope, single basin (SSSB) was selected for the study and application for rural location as it's the cheapest and most easily constructed basins [13].

The **objective** of the research is to eliminate metal from the water identifying and testing cost-effective, simple to operate and maintain and is sustainable on long runs on test field site. Also, the research investigation targeted to check the system success for arsenic-free water supply in terms of three important components, i.e., social viability and acceptance, technical and economic viability.

Methodology for solving the problem: The research used solar still technology for drinking water production, which is metal ion free and also treatment of leftover brine with natural options like plant biomass. The schematic representation for the same is given below (Fig. 2). The study consisted of four phases as described below:

- Phase I: Investigations on solar distillation unit suitable for domestic water supply, field trials for entire village.
- Phase II: Investigations for project design for application of the system in rural village with remediation and analysis per unit cost of water from the system.
- Phase III: Analysis of system in terms of sustainability parameters—social acceptance technical and economic viability.
- Phase IV: Monitoring the health targets and benefits attained by rural people.

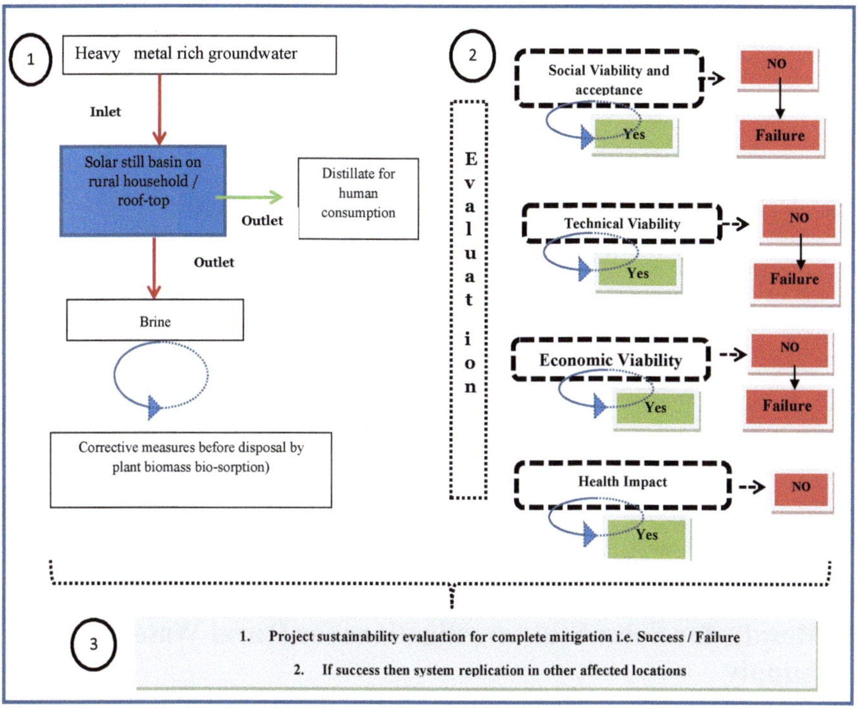

Fig. 2 Schematic representation of the concept for arsenic removal and system evaluation for sustainability

3 Experimental Set up

Figure 3 shows the unit of SSSB (single slope, single basin)-type solar still for which performance was evaluated. The unit was 1 m^2 (both length and breadth) in dimensions with a tilt angle of 8.6°. The basin was made of black fiber with a total carrying capacity of about 40 L for the basin and had a glass top ranging to 4 mm thickness. SSSB was placed in east to west direction and had both the inlet and the outlet pipes made of steel material (this helped to avoid corrosion). Different combination of sample water spiked with the variability of arsenic was added to the basin body through the inlet pipe for the 10-day study period. For the operations of the unit and experimental analysis, batch study was done, and every 10th day of outlet, distillate was collected (approximately ranged between 1.5 and 3 L) for sample analysis against different input water load.

Fig. 3 Single slope, single basin-type solar still with distillate collector [7]

4 Results for Solar Still Investigations for Rural Water Supply

This section brings forth the results of the study where inlet water quality has been compared to outlet water quality with reference to WHO drinking water standards. Referring to Fig. 4, it is evident that there exists a sharp decline in the values of TDS and chlorides both, for the outlet distillate in comparison to initial inlet water quality. TDS value was found to be lowered down to nearly 95% (i.e., from 1200–1350 range mg/L to 25–50 mg/L). Similarly, chlorides value was reduced to 96% (i.e., from 300–500 mg/L to 25–50 mg/L). No odor emissions were found in the run. The brine therefore had only 3.5–4.0% of TDS and chloride salts as leftovers. The same trend of similarity was found for dissolved arsenic, iron, nitrates, alkalinity, pH, hardness and sulfate salts.

In compliance to the drinking water standards (WHO), the level of nitrate salts should be less than 50 mg/L; for arsenic, it should be less than 0.01 mg/L and fluorides should exist as 1.5 mg/L. These were found in regulation to WHO limits. In reference to WHO standards, to make the distillate water potable and for human drinking use, some additional fluoride salts can be added as a supplement. As per WHO standard, as safe drinking water should not have coliform in any 100 ml of sample (Table 1). There were no such Coliforms found in the sample post-treatment.

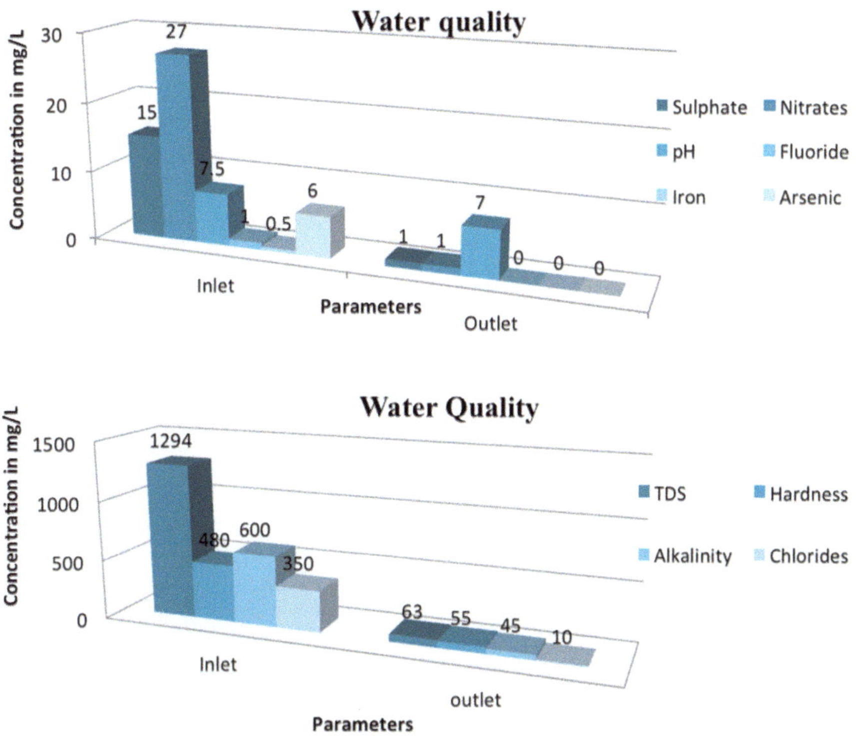

Fig. 4 Inlet and outlet water quality after distillation

5 Discussion and Conclusions

The literature has already established that the solar energy application in form of distillation using a solar still has very high treatability potential for various parameters of water to make it potable at different situations. In the study, two major conclusions can be drawn with respect to arsenic rich-groundwater—a) treated water was found to be free from pathogens and Coliforms both and b) the study results also highlighted that the selected basin has an efficiency to remove pollutant (arsenic) for more than 97%. The study results strikingly also brought forward that the other remaining parameters do meet the required drinking water standards as in accordance with WHO limit [14, 15] but with some addition of salts. This is in reference to making potable water fit for drinking; some post-treatment measure is required as the distillate water is deficient in essential salts (needed for human metabolism). Therefore, these essential salts have to be added to the distillate for drinking purpose. This would be in accordance with the current requirements as per drinking water quality standards that need 1.5 mg/L of fluoride, 240 mg/L of chlorides, etc.

The present study aimed at addressing the global problem of arsenic as a heavy metal dissolved in groundwater blocking access to potable water supply to many

Table 1 Comparison of distillate water quality with drinking water standards

S. No.	Parameter	Outlet average	WHO limits (mg/L)	Remarks
1	pH	7.14	6.5–8.0	Post treatment not required
2	TDS	45	600	Post treatment not required
3	Arsenic	<0.01	0.01	Post treatment not required
4	Alkalinity	38.3	Not defined	Post treatment not required
5	Hardness	33.8	200	Post treatment not required
6	Sulfate	0.72	250	Post treatment not required
7	Fluoride	0.02	1–1.5	Needs supplement for Fluoride salts
8	Chlorides	10.8	251	Needs supplement for Fluoride salts
9	*Coli*	NA	Should not be detectable in any 100 mL of sample	Post treatment not required
10	Iron	0.00	0.3	Post treatment not required
11	Nitrates	0.74	51	Post treatment not required

Taken from reference no. 15

vulnerable residents [16]. The research study was a contribution made for this global concern-wide spreading at an alarming rate. The study can be identified as one the most sustainable—novel concepts as (a) it's a simple concept with no requirement of skilled labor, (b) it runs on renewable energy-solar sunshine, (c) environment friendly as brine is also co-treated before release, and (d) it is sustainable on long run with less maintenance cost-effective.

The method explained in Fig. 2 clearly defines these points of credit for this unique system as a novel concept meant for human betterment. It is actually due to the ease of working and operation of the selected solar still basin SSSB (single slope, single basin) that the system is considered to be sustainable with a novel concept. The following inferences can be summed from this study:

1. Treatment offers a simple and easy to operate method for the water supply and sanitation in arsenic affected rural areas.
2. The method is independent of any requirement of electric power as it uses solar energy.
3. It does not generate any hazardous by-products and has a negligible operating cost.

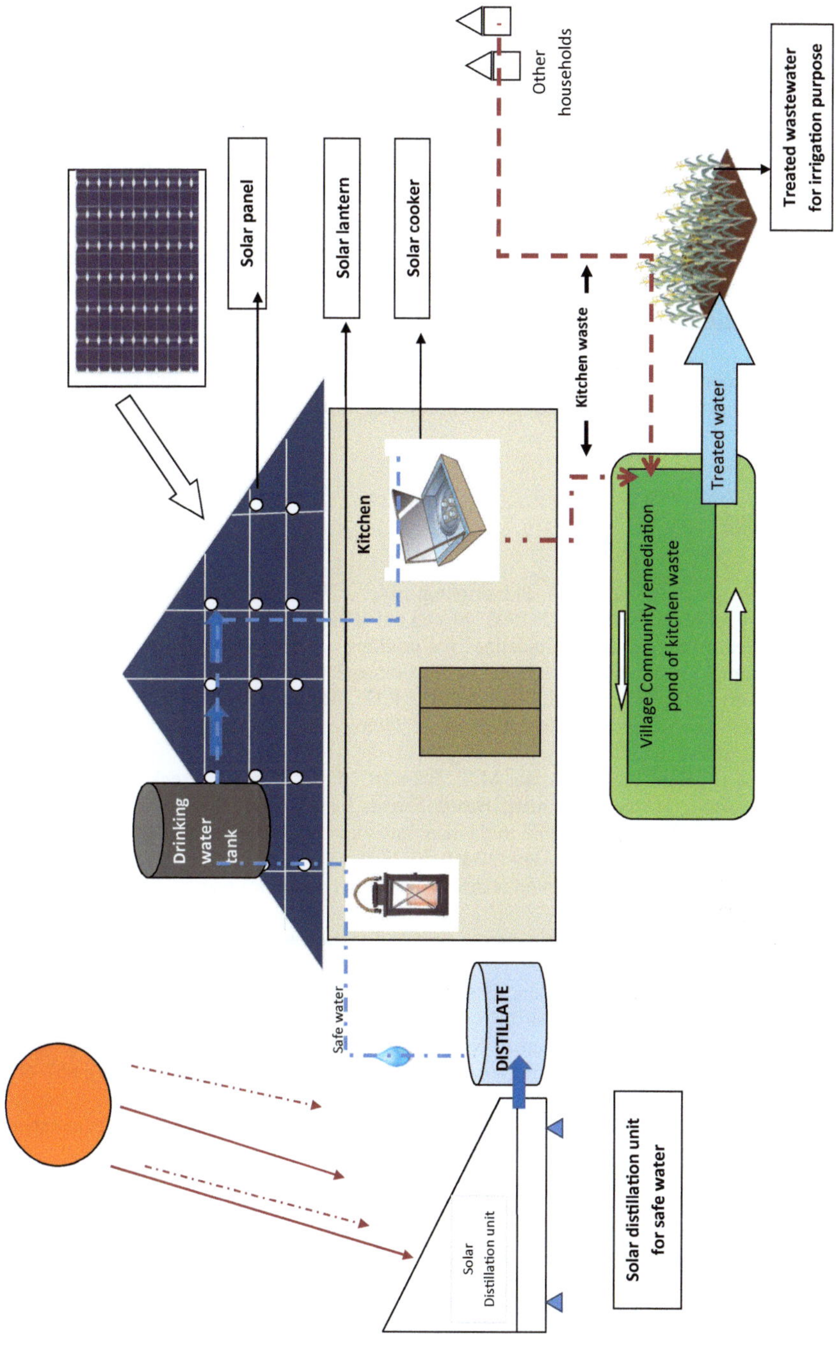

Fig. 5 Schematic for utilization of solar energy solution and distillation unit for rural household [9]

4. The experimental results show that the treated water meets WHO drinking water standards.

The co-benefits of the system are health improvement of the community, hygiene and public welfare, and a complete removal of arsenic from water ecosystem. It is actually the women with families who are required to fetch water for family use in the general rural household. Installation of solar stills on the ground or rooftops can help them save their time and runs to different locations in search for water. The time saved can be further be used by them for personal interest or education, etc. There exists ample sunlight and space in rural areas; therefore, solar distillation along with other solar-dependent energy-saving options for rural household could be a green and safe solution in upcoming time (Fig. 5).

Acknowledgements The author is grateful to the Ministry of New and Renewable Energy (MNRE New Delhi) for providing financial support as fellowship for this study.

References

1. WHO. Progress on sanitation and drinking-water—2013 update. http://apps.who.int/iris/bitstream/10665/81245/1/9789241505390_eng.pdf (2013). Accessed on 15th Jan 2016
2. World Health Organization. Guidelines for drinking—water quality (electronic resource): incorporating 1st and 2nd addenda, vol. 1, Recommendations, 3rd ed (2008)
3. Bhattacharya, P., Welch, A.H., Stollenwerk, K.G., McLaughlind, M.J., Bundschuhe, J., Panaullahf, G.: Arsenic in the environment: biology and chemistry. Sci. Total Environ. **379**(2–3), 109–120 (2007)
4. Milton, A.H., Hore, S.K., Hossain, M.Z., Rahman, M.: Bangladesh arsenic mitigation programs: lessons from the past. Emerg. Health Threats J. **5**(1), 7269 (2012)
5. Rammelt, C.F., Boes, J.: Arsenic mitigation and social mobilization in Bangladesh. Int. J. Sustain. High. Educ. **5**(3), 308–319 (2004)
6. Delyannis, E.: Historic background of desalination and renewable energies. Sol. Energy **75**(5), 357–366 (2003)
7. Tzen, E., Morris, R.: Renewable energy sources for desalination. Sol. Energy **75**, 375–379 (2003)
8. Arjunan, T.V., Aybar, H.S., Nedunchezhian, N.: Status of solar desalination in India. Renew. Sustain. Energy Rev. **13**(9), 2408–2418 (2009)
9. Avvannavar, S.M., Mani, M., Kumar, N.: An integrated assessment of suitability of domestic solar still as a viable safe water technology for India. Environ. Eng. Manage. J. **7**(6), 667–685 (2008)
10. Bouchekima, B.: A small solar desalination plant for the production of drinking water in remote arid areas of southern Algeria. Desalination **159**, 197–204 (2003)
11. Kaushal, A., Varun. Solar stills: a review. Renew. Sustain. Energy Rev. **14**(1), 446–453 (2010)
12. Tiwari, G.N., Singh, H.N., Tripathi, R.: Present status of solar distillation. Sol. Energy **75**(5), 367–373 (2003)
13. Badran, O.O., Al-Tahaineh, H.A.: The effect of coupling a flat-plate collector on the solar still productivity. Desalination **183**(1–3), 137–142 (2005)
14. Jasrotia, S., Kansal, A., Kishore, V.V.N.: Arsenic phyco-remediation by *Cladophora sp* algae and measurement of arsenic speciation and location of active absorption site using electron microscopy. Microchemical **114**, 197–202 (2014)

15. Jasrotia, S., Kansal, A., Kishore, V.V.N.: Application of solar energy for water supply and sanitation in arsenic affected rural areas a study for Kaudikasa village, India. J. Cleaner Prod. **60**(1), 102–106 (2013)
16. Jasrotia, A., Kansal, A., Sharma, M., Gupta, S.: Application of sustainable solar energy solutions for rural development—a concept for remote villages of india. In: Siddiqui, N.A., Tauseef, S.M., Bansal, K. (eds.) Advances in Health and Environment Safety. Springer Singapore (2018). ISBN: 978-981-10-7121-8

Biodegradable Plastics Production by Cyanobacteria

Ranjana Bhati

Abstract Commercial production of polyhydroxyalkanotes (PHAs), used as "green" thermoplastics in medicinal, fibre and agricultural field, is done by bacterial fermentation. The major drawback with bacterial system is high fermentation cost incurred due to expensive substrate and continuous oxygen supply. Therefore, in order to make PHAs production economically competitive, research efforts abound to exploit photoautotrophic hosts. In this context, phototrophic cyanobacteria are rising as a promising host for high PHAs accumulation in cytoplasm. Due to short generation time and minimum nutrient requirements, cyanobacterial cultivation is quite successful in wastewaters. Cyanobacteria accumulate high PHAs with various growth conditions, i.e. photoautotrophic, chemoheterotrophic with various carbon sources like glucose, fructose. It is evident from the recent studies that limiting nutrients supply and supplementation of excess carbon sources in the culture medium results in improved PHAs yield. Moreover, they grow better when fed with carbon dioxide, the major greenhouse gas. Therefore, utilization of cyanobacteria with wastewaters and carbon dioxide, for PHAs production seems highly promising as it has the dual advantage of polymer production and wastewaters recycling with photosynthetic utilization of CO_2. This chapter presents an overview on the progresses and prospects of PHAs production from cyanobacteria. Different production strategies with nitrogen and non-nitrogen fixing cyanobacteria are discussed. Material and thermal properties of the produced PHAs are also analysed and compared with commercial polymers. This opens up the possibilities for low-cost production of PHAs polymers from cyanobacteria.

Keywords Polyhydroxyalkanotes (PHAs) · Cyanobacteria

R. Bhati (✉)
Department of Microbiology, Bundelkhand University, Jhansi 284128, Uttar Pradesh, India
e-mail: ranjana.iitkgp@gmail.com

M. Khoobchandani and A. Saxena (eds.), *Biotechnology Products in Everyday Life*, EcoProduction, https://doi.org/10.1007/978-3-319-92399-4_9

1 Introduction

In the last few decades, applications of petroleum-based plastics have increased significantly worldwide after their debut in 1950s. Plastics have gained tremendous importance in every sphere of human activities and become an integral part of human's day-to-day life. Plastics are resistant to decomposition by micro-organisms (non-biodegradable), highly resistant and have long lifespan in nature [1].

To decrease the environmental impact caused due to the excessive plastic uses, a number of efforts are being made. Expensive solid waste disposal, dangerous incineration of plastic waste and reducing capacity of municipal landfills are the major concerns for bringing the world into the mainstream of eco-friendly and sustainable development. Therefore, substitution of synthetic plastics by biodegradable plastics is one of the key interests of basic and applied research. Research on biodegradable polymers is the need of the hour.

Growing concern for environment has resulted into the development of a variety of biodegradable polymers with potential as environment friendly 'green plastics' [2]. Three main classes of biodegradable plastics are:

1.1 Chemically Synthesized Plastics

Polyglycolic acid, polylactic acid, poly(ethylene oxide), polycaprolactone and polyvinyl alcohol are included in this category. These are biodegradable. They are not commercially viable as alternative for petro-chemical plastics as they do not possess all the properties similar to plastics [2].

1.2 Starch-Based Biodegradable Plastics

In this type, starch is added to hold together the fragments of plastics (e.g. starch-polyethylene). When starch-linked plastics are discarded into landfills, soil micro-organisms degrade the starch easily, thus releasing the polymer fragments. Starch is degraded by bacteria, but polyethylene fragments remain non-biodegradable, which remain in environment [3].

1.3 Completely Biodegradable Plastics

This type of plastics is completely utilized by micro-organisms. These are the only 100% biodegradable polymers. Polyhydroxyalkanoates (PHAs), polylactides (PLA), polysaccharides, aliphatic polyesters and co-polymers are included in this category.

Among all, the microbially produced PHAs offer important contributions as 'bioplastics'. Bioplastics are plastics produced from renewable resources and completely biodegradable.

2 Polyhydroxyalkanoates (PHAs)

PHAs constitute a family of polyesters of wide range of different hydroxyalkanoic acids (HAs). PHAs are synthesized by numerous micro-organisms as intracellular carbon and energy compound under limiting nutrition conditions (such as phosphorus and nitrogen) with presence of excess carbon sources [4]. Physical properties of PHAs are similar to various synthetic plastics like polypropylene, and therefore can be used as substitute for conventional plastics [5].

PHAs are derived from the renewable sources and also completely degraded to carbon dioxide and water under aerobic conditions and to methane under anaerobic conditions by micro-organisms in soil, sea, lake water and sewage [2, 5]. PHAs accumulation by micro-organisms can be stimulated with numerous growth conditions including nutrient limitation and intermittent substrate feeding [4, 6, 7]. More than 300 species of micro-organisms are known to synthesize and accumulate PHAs [8]. Research and commercial interest in PHAs have increased recently since they are biocompatible [9] and have immense scope for medical applications [10]. Use of PHAs for packaging and other goods can help in preserving the environment by reducing the pollution and conserve oil reserves. PHAs also have wide applications for tissue engineering, surgical implants, drug delivery and other specialized fields [11–13]. PHAs are also used as starting material to produce enantiomerically pure drugs and chemicals [14].

2.1 *Structure of PHAs*

Based on the number of carbon atoms in the chain, PHAs can be divided into three classes, i.e. short-chain length (SCL), consisting of 3–5 carbon atoms, medium-chain length (MCL), consisting of 6–14 carbon atoms and long-chain length (LCL), consisting of more than 14 carbon atoms [2, 4, 15, 16] (Fig. 1).
For example:

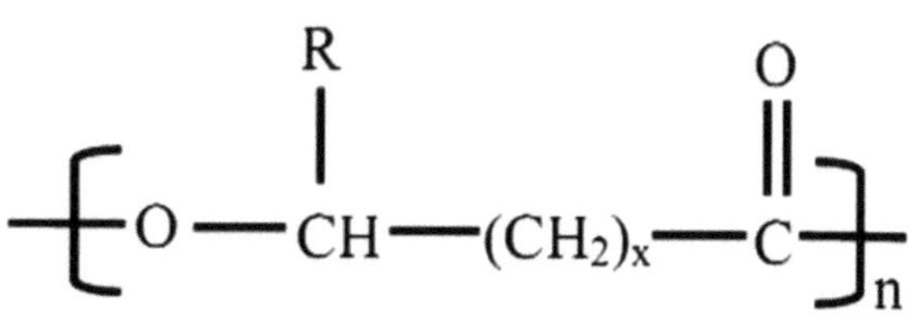

Fig. 1 Structure of polyhydroxyalkanoates [4, 15]

$x = 1$	R = hydrogen	Poly(3-hydroxypropionate)
	methyl	Poly(3-hydroxybutyrtate)
	ethyl	Poly(3-hydroxyvalerate)
$x = 2$	R = hydrogen	Poly(4-hydroxybutyarte)
$x = 3$	R = hydrogen	Poly(5-hydroxyvalerate)

2.2 *Poly(3-Hydroxybutyrate-*co*-3-Hydroxyvalerate) [P(3HB-*co*-3HV)] Co-polymer*

Among the 150 types of PHA monomers studied so far, the homopolymer of hydroxybutyrate, i.e. poly(3-hydroxybutyrate) (PHB) is most commonly found in prokaryotes including cyanobacteria and also the best-known characterized member of PHA family [17]. However, PHB has poor material properties such as high crystallinity, brittleness and low mechanical strength, thus its application is limited. Moreover, its high melting temperature, i.e. about 170 °C, makes its processing a difficult task [6]. Initial research developments were therefore directed for making PHAs that have superior material properties and easier to process. Poly(3-hydroxybutyrate-*co*-3-hydroxyvalerate) or P(3HB-*co*-3HV) co-polymer is a co-polymer of PHB containing random segments of hydroxyvalerate and is found to have superior material properties as compared to PHB. P(3HB-*co*-3HV) co-polymer is less stiff and tougher. Incorporation of 3-hydroxyvalerate, i.e. 3(HV) units reduces the crystallinity and melting temperature (T_m) of PHB homopolymer and, gives better flexibility and tensile strength to the polymer [6]. This is advantageous as low T_m widens the size of processing window as compared to PHB. P(3HB-*co*-3HV) co-polymer can be used to prepare films with excellent water and gas barrier properties and can be processed at a lower temperature while retaining most of the other mechanical properties of PHB homopolymer [18]. Therefore, the commercial production of reformative P(3HB-*co*-3HV) co-polymer has been exploited for a stronger and more flexible plastic (Fig. 2).

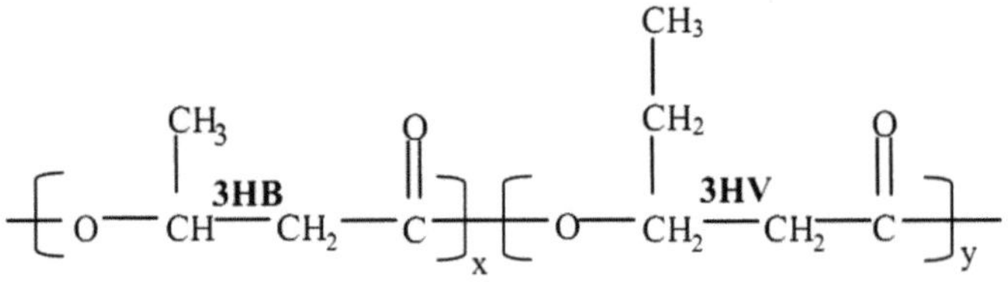

Fig. 2 P(3HB-*co*-3HV) co-polymer [4, 15]

Table 1 PHA manufacturers worldwide [19]

Trade name	Polymer	Manufacturer
Biogreen	PHB	Mitsubishi Gas Chemical Company Inc. (Japan)
Biocycle®	PHB	PHB Industrial Company (Brazil)
Mirel™	PHB	Telles (USA)
Biopol	PHBV	Metabolix (Cambridge, MA, USA)
Biomer®	PHB and PHBV	Biomer Inc. (Germany)
Enmat®	PHB and PHBV	Tianan Biologic, Ningbo (China)
Nodax™	PHBH	Procter & Gamble (USA)
		Lianyi Biotech (China)
Kaneka PHBH	PHBH	Kaneka Corporation (Japan)
Green Bio	P(3HB-*co*-4HB)	Tianjin Gree Bio-Science Co. (China)

2.3 Applications for PHAs

Bioplastics have been used in variety of applications since the first industrial production of PHB by W.R. Grace & Co., New York, U.S. in 1959 [19]. They are widely used for the fabrication of bottles, fibres and several products of commercial, agricultural and packaging interest [16]. In response to the needs of tissue engineering, PHAs are used in medical applications as absorbable polymer and tissue scaffold. The commercial applications of PHAs are in agriculture, pharmaceuticals and medical fields [17]. Biodegradability and hydrophobicity makes PHAs suitable for use in packaging industry and hygiene products [12].

2.4 Commercially Available PHAs

Commercial PHBV production was done by Imperial Chemical Industries Ltd. (ICI/Zeneca BioProducts, Bellingham, UK) with the trade name of Biopol™ in 1970. Then the technology was taken by Monsanto (USA) and then by Metabolix [19]. Soon worldwide many companies commercialized PHAs solely or in collaboration with research institutes as detailed in Table 1.

3 P(3HB-*co*-3HV) Co-polymer Accumulation in Bacteria and Commercial Drawbacks

A large number of bacteria (over 300 species, *Bacillus* sp., *C. necator*, *Methylobacterium* sp., *Pseudomonas* sp., etc.) have the metabolic capacity to synthesize PHAs. In heterotrophic bacteria, maximum PHB accumulation amounted up to 80% dry cell weight has been produced with fermentation [20]. P(3HB-*co*-3HV) co-polymer synthesis in some bacterial species are reported with supplementation of appropriate carbon sources such as acetate, glucose, propionate, glycerol and valerate. Bacterial PHA production can be seen in the extensive reviews of Khanna and Srivastava [2], Anderson and Dawes [4], Braunegg et al. [5], Lee [6], Lee et al. [8], Philip et al. [14], Reddy et al. [20], Doi [21], Steinbüchel [22], Sudesh et al. [23] and Suriyamongkol et al. [24].

Despite the various researches from the past 75 years, PHAs seem to be far reached from commercial production. The major obstacle for wide acceptance of PHAs, over the replacement of synthetic plastics, is their cost difference (€3.5–5.0/kg for PHA compared with €1/kg for polypropylene) [19]. PHA is commercially produced by pure cultures of micro-organisms, such as *C. necator* or recombinant *E. coli*, and the main cost of production by bacteria is primarily due to expensive carbon substrate, oxygen supply and need for sterile operation during bacterial fermentation process [6]. Economic analysis of PHAs production process suggested that 50% of the total production cost is contributed mainly by the cost of carbon substrate [25]. Consequently, in order to make PHAs production economically competitive, research efforts abound to exploit photoautotrophic organisms or other low-cost substrates.

4 Attempts for PHAs Production in Transgenic Plants

Photoautotrophic PHAs accumulation in agricultural crops is another potential alterative for large-scale and low-cost production [26, 27]. PHA production in plant system is considerably less expensive when compared with bacteria because plants require water, soil nutrients, CO_2 and sunlight. Plants photosynthetically fix CO_2 and water to accumulate biopolymer, which are degraded back to CO_2 and water after disposal [24]. PHA synthesis in crops economically increases the value of crops [26, 28].

First attempt of PHB production in plants was carried out with model plant, *Arabidopsis thaliana*, and transformed with *Agrobacterium tumefaciens*. Intracellular PHB accumulation reached up to 0.1% of the shoot dry weight. Moreover, a negative impact was found on the growth of the plant [29]. A negative correlation between PHB accumulation and plant growth was observed in plants accumulating PHB more than 5% (dcw). With these experiments, it was concluded that high level of PHB synthesis in the plastids beyond some critical point had adverse effect on chloroplast function [30].

A landmark in plant PHA production was achieved by synthesis of P(3HB-*co*-3HV) co-polymer in *A. thaliana* and rapeseed *(Brassica napus)* [31]. Expression of all the genes for P(3HB-*co*-3HV) co-polymer in *A. thaliana* with CaMV35S promoter resulted in accumulation between 0.2 and 0.8% (dcw) in shoots, with 4–17 mol% hydroxyvalerate content. P(3HB-*co*-3HV) co-polymer accumulation up to 2.3% (dcw) with 6.5 mol% hydroxyvalerate content was reported for seed-specific expression of the same pathway in leucoplast of rape embryos [31]. P(3HB-*co*-3HV) co-polymer accumulation was consistently lower compared to PHB in both *Arabidopsis* and rapeseed, although the same *C. necator* gene constructs were used in these different transgenic plants. Overexpression of the threonine deaminase and synthesis of propionyl-CoA indicates a negative metabolic effect which eventually affects the PHAs accumulation [30]. These studies indicates the major challenges involved in modifying several independent pathways, as in the case of P(3HB-*co*-3HV) co-polymer with fatty acid and amino acid biosynthetic pathways, and to supply different monomers for the synthesis of PHA co-polymers. Therefore, low expression level, retarded growth and difficulties in isolating the polymer are the major limitations in plant-based polymer production [30].

5 Cyanobacteria as Production Hosts for PHAs

Polyhydroxyalkanoates from cyanobacteria have been an important subject of research since last five decades. Cyanobacteria (blue-green algae) are photoautotrophic gram-negative prokaryotes that carry out oxygenic photosynthesis for PHAs accumulation. Cyanobacteria are one of the oldest life forms on earth and simplest known photoautotrophic organisms having characteristics of both algae and bacteria. They carry out oxygenic photosynthesis with possession of chlorophyll *a* and phycobiliproteins as photosynthetic pigments [32]. They need simple inorganic nutrients such as calcium, magnesium, potassium, phosphate, nitrate (not in case of nitrogen-fixers) and sodium as macronutrient, and B, Co, Cu, Fe, Mn, Mo and Zn as micronutrients for their growth and multiplication [33]. Further, wastewater can be used for their cultivation due to their ability to utilize inorganic nitrogen and phosphorus for their growth. Photoautotrophic PHAs production in cyanobacteria may overcome the expense incurred due to feeding of organic carbon and oxygen supply during bacterial fermentation. Therefore, utilization of cyanobacteria for PHAs production with carbon dioxide, and wastewaters seem highly encouraging as it has the dual advantage of polymer production and recycling the wastewaters while photosynthetically utilizing CO_2.

PHAs accumulation in cyanobacteria has been the subject of intense research, after Carr [34] demonstrated PHB accumulation in *Chlorogloea fritschii* by chemical analysis. Further Jensen and Sicko [35] demonstrated PHB inclusion bodies with the same cyanobacterium by electron microscopy. Since then, cyanobacterial PHAs accumulation has been studied by many researchers, but the contents are found to be very low and amount less than 10% (dcw) with photoautotrophic growth conditions [36–39].

A landmark in cyanobacterial PHA research was achieved when Nishioka et al. [40] found PHB accumulation up to 55% (dcw) under phosphate-limited condition in a unicellular cyanobacterium, *Synechococcus* sp. MA19, isolated from the volcanic rock of Japan. Further, N_2-fixing cyanobacterium *Nostoc muscorum* was reported with PHB accumulation up to 46% with chemoheterotrophy under phosphate limitation [41]. This N_2-fixing cyanobacterium was also able to synthesize the P(3HB-*co*-3HV) co-polymer up to 31% under acetate and propionate-supplemented condition [42]. More recently, high PHB accumulation up to 85% (dcw) was reported for cyanobacterium, *Aulosira fertilissima* CCC 444 with 0.28% acetate, 0.26% citrate and 5.58 mg l^{-1} K_2HPO_4-supplemented condition [43]. Further, studies with the same strain reported the maximum P(3HB-*co*-3HV) co-polymer accumulation up to 77% (dcw) with fructose and valerate supplementation [44]. In *Nostoc muscorum* Agardh P(3HB-*co*-3HV) co-polymer production increased up to 58 and 60% (dcw) with 0.4% acetate and 0.4% valerate supplementation under phosphorus and nitrogen deficiencies, respectively [45]. Further rise in co-polymer content was recorded up to 78% (dcw) with optimized condition (0.28% acetate, 0.38% glucose and 0.3% valerate) and N-deficiency after 7 days of incubation [46]. Table 2 presents the some cyanobacterial strains with high PHAs accumulation potential.

6 Mechanical and Thermal Properties of PHAs

Physical properties of PHAs are significant in determining the polymer application and suitability in different fields. For producing a range of different thermoplastics with varying degrees of toughness and flexibility, the variations in 3HV content in P(3HB-*co*-3HV) co-polymer are desirable. A comparison of mechanical and thermal properties of PHB homopolymer and P(3HB-*co*-3HV) co-polymer with different 3HV fractions (mol%) obtained from cyanobacteria (*Aulosira fertilissima* and *Nostoc muscorum*) with commercial co-polymers is listed in Table 3.

Overall, the material properties of the films isolated from cyanobacterial sources were comparable with the commercial co-polymers and advocate its potential application in various fields. Thus, cyanobacteria can be considered as a suitable source for P(3HB-*co*-3HV) co-polymer production for agricultural as well as pharmaceutical applications.

7 Conclusion and Future Perspectives

Polyhydroxyalkanotes are prime focus of environmental, agricultural and biomedical sciences nowadays. Various studies show that cyanobacteria could act as alternative phototrophic host for polymer production and intensive studies are required to

Table 2 Reports on increased PHAs accumulation in some cyanobacterial species

Name of the cyanobacterium	Culture condition	Polymer type (% dcw)	References
Synechococcus sp. MA19	Photoautotrophy nitrogen-starved	PHB (27%)	[47]
Anabaena cylindrical 10 C	Acetate + propionate	P(3HB-*co*-3HV) (2%)	[36]
Synechococcus sp. MA19	Photoautotrophy phosphate limitation	PHB (55%)	[40]
Nostoc muscorum	Chemoheterotrophy (acetate)	PHB (43%)	[33]
Synechocystis sp. PCC 6803	Mixotrophy (glucose + acetate) + phosphate-deficiency	PHB (29%)	[37]
Synechocystis sp. PCC 6803	Mixotrophy (acetate + fructose) + gas exchange limitation + phosphate-deficiency	PHB (38%)	[48]
Nostoc muscorum	Chemoheterotrophy (acetate + glucose) + phosphate limitation	PHB (46%)	[41]
	Acetate (0.11%) + propionate (0.08%)	P(3HB-*co*-3HV) (31%)	[42]
Synechocystis sp. UNIWG	Mixotrophy (acetate) + nitrogen-starved	PHB (14%)	[49]
Nostoc muscorum	Phosphate-starved + (aeration + CO_2)	PHB (22%)	[50]
Nostoc muscorum Agardh	Acetate + valerate + nitrogen-starved	P(3HB-*co*-3HV) (60%)	[45]
Aulosira fertilissima CCC 444	Citrate + acetate and K_2HPO_4	PHB (85%)	[43]
	Fructose + valerate	P(3HB-*co*-3HV) (77%)	[44]
	Acetate + gas exchange limitation	PHB (49%)	[51]
Nostoc muscorum Agardh	Optimized condition	P(3HB-*co*-3HV) (78%)	[46]
	CO_2 + poultry litter	P(3HB-*co*-3HV) (65%)	[52]
Synechocystis sp. PCC 6714	Photoautotrophy phosphate limitation + nitrogen limitation	PHB (16.4%)	[53]

Table 3 Properties of commercial PHAs and PHAs films produced by cyanobacteria

Property	Commercial PHAs			PHAs *from Aulosira fertilissima*		PHAs from *Nostoc muscorum* Agardh		
	PHB	P(3HB-*co*-3HV) (mol fraction)		P(3HB-*co*-3HV) (mol fraction)		PHB	P(3HB-*co*-3HV) (mol fraction)	
	100:0	86:14	75:25	100:0	75:25	100:0	84:16	78:22
T_m (°C)	180	150	137	174	155	176	154	148.8
T_g (°C)	0–5	−4	−6	0.6	−5.5	0.8	−2.2	−4.3
X_c (%)	60–80	59.6	40	60.7	46.1	62.4	44.7	39.5
Young's modulus (GPa)	3.5	1.5	0.7	3.4	0.6	3.9	1.2	0.8
Tensile strength (MPa)	40	35	30	37.6	18.1	32.4	26.5	21.5
Elongation to break (%)	5	20	nr	4.9	87.2	4.7	72.1	85.1
Reference	[17, 54–56]			[44]		[45, 46, 57]		

enhance PHA productivity in cyanobacteria. Only few cyanobacteria are reported to accumulate PHAs at par with bacterial species. However, bacterial PHA productivity is many folds higher when compared with cyanobacterial productivity. Cyanobacterial productivity should be further improved for cost-effective PHAs production. PHB and P(3HB-*co*-3HV) co-polymers with wide monomer composition can be efficiently produced by cyanobacteria with changing substrate concentration in growth medium.

Research efforts are needed to develop novel strategies for improved cyanobacterial growth, development and exploitation of better strains for cost-effective PHAs production in near future. Extensive studies are required for improvements of downstream processing for better utilization of cyanobacterial biomass to produce eco-friendly bioplastics.

Moreover, cyanobacteria are potent host for producing biopolymers by CO_2 fixation and wastes utilization. In the current scenario, this is no doubt one of the powerful, economic and eco-friendly processes of CO_2 mitigation and waste recycling for biomass and PHAs production. At present, only few reports are available in this area and intensive research is essential in near future for resourceful utilization of cyanobacterial biomass to produce eco-friendly affordable bioplastic. Therefore, intense researches are required at pilot-scale for commercial production of cyanoplastics.

Conflict of Interest Author declares no conflict of interest.

References

1. Loo, C.Y., Sudesh, K.: Polyhydroxyalkanoates: biobased microbial plastics and their properties. Malasian Polym. J. **3**(2), 31–57 (2007)
2. Khanna, S., Srivastava, A.K.: Recent advances in microbial polyhydroxyalkanoates. Process. Biochem. **40**(2), 607–619 (2005)
3. Johnstone, B.: A throw away answer. Far Eastern Econ. Rev. **147**(6), 62–63 (1990)
4. Anderson, A.J., Dawes, E.A.: Occurrence, metabolism, metabolic role, and industrial uses of bacterial polyhydroxyalkanoates. Microbiol. Rev. **54**(4), 450–472 (1990)
5. Braunegg, G., Gilles, L., Klaus, F.: Polyhydroxyalkanoates, biopolyesters from renewable resources: physiological and engineering aspects. J. Biotechnol. **65**(2), 127–161 (1998)
6. Lee, S.Y.: Bacterial polyhydroxyalkanoates. Biotechnol. Bioeng. **49**(1), 1–4 (1996)
7. Reis, M.A.M., Serafim, L.S., Lemos, P.C., Ramos, A.M., Aguiar, F.R., Van Loosdrecht, M.C.M.: Production of polyhydroxyalkanoates by mixed microbial cultures. Bioprocess. Biosyst. Eng. **25**(6), 377–385 (2003)
8. Lee, S.Y., Choi, J., Wong, H.W.: Recent advances in polyhydroxyalkanoate production by bacterial fermentation: mini-review. Int. J. Biol. Macromol. **25**(1–3), 31–36 (1999)
9. Brandl, H., Gross, R.A., Lenz, R.W., Fuller, R.C.: Plastic from bacteria and for bacteria: poly(beta-hydroxyalkanoates) as natural, biocompatible and biodegradable polyesters. Adv. Biochem. Eng./Biotechnol. **41**, 77–93 (1990)
10. Zinn, M., Witholt, B., Egli, T.: Occurrence, synthesis and medical application of bacterial polyhydroxyalkanoate. Adv. Drug. Deliv. Rev. **53**(1), 5–21 (2001)
11. Williams, S.F., Martin, D.P., Horowitz, D.M., Peoples, O.P.: PHA applications: addressing the price performance issue I. Tissue engineering. Int. J. Biol Macromol. **1999**, 25(1–3), 111–121
12. Chen, G.Q., Wu, Q.: The application of polyhydroxyalkanoates as tissue engineering materials. Biomaterials **26**(33), 6565–6578 (2005)
13. Thomson, N., Roy, I., Summers, D., Sivaniah, E.: In vitro production of polyhydroxyalkanoates: achievements and applications. J. Chem. Technol. Biotechnol. **85**(6), 760–767 (2010)
14. Philip, S., Keshavarz, T., Roy, I.: Polyhydroxyalkanoates: biodegradable polymers with a range of applications. J. Chem. Technol. Biotechnol. **82**, 233–247 (2007)
15. Steinbüchel, A.: Biodegradable plastics. Curr. Opin. Biotechnol. **3**(3), 291–297 (1992)
16. Steinbüchel, A., Füchtenbusch, B.: Bacterial and other biological systems for polyester production. Trends Biotechnol. **16**(10), 419–427 (1998)
17. Vincenzini, M., De Philippis, R.: Polyhydroxyalkanoates. In: Cohen, Z. (ed.) Chemicals from Microalgae, pp. 292–312. Taylor and Francis Inc., London, USA (1999). ISBN 9780748405152
18. Luzier, W.D.: Materials derived from biomass/biodegradable material. Proc. Natl. Acad. Sci. U. S. A. **89**(3), 839–842 (1992)
19. Chanprateep, S.: Current trends in biodegradable polyhydroxyalkanoates. J. Biosci. Bioeng. **110**(6), 621–632 (2010)
20. Reddy, C.S.K., Ghai, R., Kalia, V.C.: Polyhydroxyalkanoates: an overview. Bioresour. Technol. **87**(2), 137–146 (2003)
21. Doi, Y.: Microbial Polyesters, 1st edn., p. 166. VCH Publishers, New York, USA (1990). ISBN-13:978-0471187325
22. Steinbüchel, A.: PHB and other polyhydroxyalkanoic acids. In: Doi, Y., Fukuda, K. (eds.) Biodegradable Plastics and Polymers, pp. 362–364. Elsevier, New York, USA (1996). ISBN 9781483290409
23. Sudesh, K., Bhubalan, K., Chuah, J.A., Kek, Y.K., Kamilah, H., Sridewi, N., Lee, Y.F.: Synthesis of polyhydroxyalkanoate from palm oil and some new applications. Appl. Microbiol. Biotechnol. **89**(5), 1373–1386 (2011)
24. Suriyamongkol, P., Weselake, R., Narine, S., Moloney, M., Shah, S.: Biotechnological approaches for the production of polyhydroxyalkanoates in microorganisms and plants—A review. Biotechnol. Adv. **25**(2), 148–175 (2007)

25. Solaiman, D.K.Y., Ashby, R.D., Foglia, T.A., Marmer, W.N.: Conversion of agricultural feedstock and coproducts into poly(hydroxyalkanoates). Appl. Microbiol. Biotechnol. **71**(6), 783–789 (2006)
26. Poirier, Y.: Production of new polymeric compounds in plants. Curr. Opin. Biotechnol. **10**(2), 181–185 (1999)
27. Poirier, Y.: Polyhydroxyalkanoate synthesis in plants as a tool for biotechnology and basic studies of lipid metabolism. Prog. Lipid Res. **41**(2), 135–155 (2002)
28. Somerville, C.R., Bonetta, D.: Plants as factories for technical materials. Plant Physiol. **125**(1), 168–171 (2001)
29. Poirier, Y., Dennis, D.E., Klomparens, K., Somerville, C.: Polyhydroxybutyrate, a biodegradable thermoplastic, produced in transgenic plants. Science **256**(5056), 520–523 (1992)
30. Rezzonico, E., Moire, L., Poirier, Y.: Polymers of 3-hydroxyacids in plants. Phytochem. Rev. **1**(1), 87–92 (2002)
31. Slater, S., Mitsky, T.A., Houmiel, K.L., Hao, M., Rieser, S.E., Taylor, N.B., Tran, M., Valentin, H.E., Rodriguez, D.J., Stone, D.A., Padgette, S.R., Kishore, G., Gruys, K.J.: Metabolic engineering of *Arabidopsis* and *Brassica* for poly(3-hydroxybutyrate-*co*-3-hydroxyvalerate) copolymer production. Nat. Biotechnol. **17**(10), 1011–1016 (1999)
32. Stal, L.J., Moezelaar, R.: Fermentation in cyanobacteria. FEMS Microbiol. Rev. **21**(2), 179–211 (1997)
33. Sharma, L., Mallick, N.: Accumulation of poly-β-hydroyxybutyrate in *Nostoc muscorum*: regulation by pH, light-dark cycles, N and P status and carbon sources. Bioresour. Technol. **96**(11), 1304–1310 (2005)
34. Carr, N.G.: The occurrence of poly-β-hydroxybutyrate in the blue-green alga, *Chlorogloea fritschii*. Biochim. Biophys. Acta **120**(2), 308–310 (1966)
35. Jensen, T.E., Sicko, L.M.: The fine structure of *Chlorogloea fritschii* cultured in sodium acetate enriched medium. Cytologia **38**(3), 381–391 (1971)
36. Lama, L., Nicolaus, B., Calandrelli, V., Maria, M.C., Romano, I., Gambacorta, A.: Effect of growth conditions on endo- and exopolymer biosynthesis in *Anabaena cylindrica* 10 C. Phytochemistry **42**(3), 655–659 (1996)
37. Panda, B., Jain, P., Sharma, L., Mallick, N.: Optimization of cultural and nutritional conditions for accumulation of poly-β-hydroxybutyrate in *Synechocystis* sp. PCC 6803. Bioresour. Technol. **97**(11), 1296–1301 (2006)
38. Shrivastav, A., Mishra, S.K., Mishra, S.: Polyhydroxyalkanoates synthesis by *Spirulina subsalsa* from Gujarat coast of India. Int. J. Biol. Macromol. **46**(2), 255–260 (2010)
39. Vincenzini, M., Sili, C., De Philippis, R., Ena, A., Materassi, R.: Occurrence of poly-β-hydroxybutrate in *Spirulina* species. J. Bacteriol. **172**(5), 2791–2792 (1990)
40. Nishioka, M., Nakai, K., Miyake, M., Asada, Y., Taya, M.: Production of poly-β-hydroyxybutyrate by thermophilic cyanobacterium, *Synechococcus sp.* MA19, under phosphate limitation. Biotechnol. Lett. **23**(14), 1095–1099 (2001)
41. Sharma, L., Singh, A.K., Panda, B., Mallick, N.: Process optimization for poly-β-hydroxybutyrate production in a nitrogen fixing cyanobacterium, *Nostoc muscorum* using response surface methodology. Bioresour. Technol. **98**(5), 987–993 (2007)
42. Mallick, N., Gupta, S., Panda, B., Sen, R.: Process optimization for poly(3-hydroxybutyrate-*co*-3-hydroxyvalerate)co-polymer production by *Nostoc muscorum*. Biochem. Eng. J. **37**(2), 125–130 (2007)
43. Samantaray, S., Mallick, N.: Production and characterization of poly-β-hydroxybutyrate (PHB) polymer from *Aulosira fertilissima*. J. Appl. Phycol. **24**(4), 803–814 (2012)
44. Samantaray, S., Mallick, N.: Production of poly(3-hydroxybutyrateco-3-hydroxyvalerate) co-polymer by the diazotrophic cyanobacterium *Aulosira fertilissima* CCC 444. J. Appl. Phycol. **26**(1), 237–245 (2014)
45. Bhati, R., Mallick, N.: Production and characterization of poly(3-hydroxybutyrate-*co*-3-hydroxyvalerate) co-polymer by a N_2-fixingcyanobacterium, *Nostoc muscorum Agardh*. J. Chem. Technol. Biotechnol. **87**(4), 505–512 (2012)

46. Bhati, R., Mallick, N.: Poly(3-hydroxybutyrate-*co*-3-hydroxyvalerate) copolymer production by the diazotrophic cyanobacterium *Nostoc muscorum* Agardh: process optimization and polymer characterization. Algal Res. **7**, 78–85 (2015)
47. Miyake, M., Erata, M., Asada, Y.A.: Thermophilic cyanobacterium, *Synechococcus sp.* MA19, capable of accumulating poly-β-hydroxybutyrate. J Ferment Bioeng. **82**(5), 512–514 (1996)
48. Panda, B., Mallick, N.: Enhanced poly-β-hydroxybutyrate accumulation in a unicellular cyanobacterium, *Synechocystis* sp. PCC 6803. Lett. Appl. Microbiol. **44**(2), 194–198 (2007)
49. Toh, P.S.Y., Jau, M.H., Yew, S.P., Abed, R.M.M., Sudesh, K.: Comparison of polyhydroxyalkanoates biosynthesis, mobilization and the effects on cellular morphology in *Spirulina platensis* and *Synechocystis* sp. UNIWG. J. Biosci. **19**(2), 21–38 (2008)
50. Haase, S.M., Huchzermeyer, B., Rath, T.: PHB accumulation in *Nostoc muscorum* under different carbon stress situations. J. Appl. Phycol. **24**(2), 157–162 (2012)
51. Samantaray, S., Mallick, N.: Impact of various stress conditions on poly-β-hydroxybutyrate (PHB) accumulation in *Aulosira fertilissima* CCC 444. Curr. Biotechnol. **4**(3), 366–372 (2015)
52. Bhati, R., Mallick, N.: Carbon dioxide and poultry waste utilization for production of polyhydroxyalkanoate biopolymers by *Nostoc muscorum* Agardh: a sustainable approach. J. Appl. Phycol. **28**(1), 161–168 (2016)
53. Kamravamanesh, D., Pflugl, S., Nischkauer, W., Limbeck, A., Lackner, A., Herwig, C.: Phtosynthetic poly-β-hydroxybutyrate accumulation in unicellular cyanobacterium, *Synechocystis* sp. PCC 6714. AMB Express **7**(1), 143 (2017)
54. Galego, N., Miguens, F.C., Sanchez, R.: Physical and functional characterization of PHA_{SCL} membranes. Polymer **43**(10), 3109–3114 (2002)
55. Ojumu, T.V., Yu, J., Solomon, B.O.: Production of polyhydroxyalkanoates, a bacterial biodegradable polymer. Afr. J. Biotechnol. **3**(1), 18–24 (2004)
56. Sankhla, I.S., Bhati, R., Singh, A.K., Mallick, N.: Poly(3-hydroxybutyrate-*co*-3-hydroxyvalerate) co-polymer production from a local isolate, *Brevibacillus invocatus* MTCC 9039. Bioresour. Technol. **101**(6), 1947–1953 (2010)
57. Bhati, R., Samantaray, S., Sharma, L., Mallick, N.: Poly-β-hydroxybutyrate accumulation in cyanobacteria under photoautotrophy. Biotechnol. J. **5**(11), 1181–1185 (2010)

Part V
Yielding More (Agricultural and Industrial)

Global Status of Genetically Modified Crops and Its Commercialization

Gulshan Chaudhary and Sumit Kumar Singh

Abstract From last 20 years, commercialization of biotechnological crops confirms that the genetically modified crops have conveyed a significant agronomic, ecological, monetary, social, and health advantages to ranchers, and along this, it witnessed increased numbers of consumers. The regular appropriation of GM crops reflects the generous numerous advantages for both minor and large-scale ranchers in developed and in developing countries. The principle point of GM crops is to accomplish food security or to improve nutrition, enhance sustenance, and advance manageable farming. In 2016, from 26 countries, 19 are developing and 7 are industrial countries and these countries established a biotech crops but according to International Service for the Acquisition of Agri-biotech Applications (ISAAA) top ten countries including both developed (USA and Canada) and developing (Argentina, Brazil, China, India, Pakistan, Paraguay, South Africa, and Uruguay) countries which are growing biotech crops in 1.0 million hectors of land. Secondly, in 2016 biotech crops are globally increased from 179.7 million hectares to 185.1 million hectares, and an increase of 3% is equal to 5.4 million hectares. As per the study in 2015 global economic benefit 15.4 billion US$ (7.5 billion US$ was from developing and 7.9 billion US$ was from industrial countries).

Keywords Genetically Modified crops · Biotech crops · Developed countries
Low human development index (HDI)

G. Chaudhary (✉)
The Energy and Resources Institute, India Habitate Centre, Lodhi Road,
New Delhi 110003, India
e-mail: chaudhary84@gmail.com

S. K. Singh
Grain Quality Lab, Department of Genetics, Indian Agriculture Research Institute,
New Delhi 110013, India

M. Khoobchandani and A. Saxena (eds.), *Biotechnology Products in Everyday Life*,
EcoProduction, https://doi.org/10.1007/978-3-319-92399-4_10

1 Introduction

Since 1996, the work on biotech crops is been carried out and the survey points toward the fact that biotech crops are important with respect to agronomics, environment, economics, health, and socially beneficial to farmers. The frequent adoption of biotech crops is a proof that both hefty and trivial agrarians in industrialized and emerging countries are reaping the benefits of commercially grown biotech crops [1, 2].

It has been recorded that in the past 20 years, around 2 billion hectares of biotech crops have been industrially developed. It contained 0.1 billion hectares of biotech canola, 0.3 billion hectares of biotech cotton, 0.6 million hectares of biotech maize, and 1.0 billion hectares of biotech soybean. It has been noticed that biotech products give 7.4 billion peoples sustenance and accommodation. On the basis of statistic data, demand for food is constantly expanding day by day. It is projected to be 9.9 billion of every 2050 and 12.3 billion out of 2100, which is a challenging assignment. As per estimates, food production needs to be increased by 50–70% to meet the world's requirement for food. Moreover, it has to be achieved with deteriorating land and water resources along with the environmental and agrarian challenges because of climate change [3].

The worldwide scientific group endorses the possibility of protected and economical approaches. It is to use the top of conventional crop technology or biotechnology. For example, it can be a combination of well-adapted agronomically desirable high-yielding germplasm and GM/non-GM traits. This will help in achieving sustainable intensification of crop throughput of cropland globally (1.5 billion hectares) [4].

In around 30 countries, more than 18 million farmers (90% small/poor farmers) have benefitted from biotech crops. In the past 20 years, they have witnessed the following positive impact:

- Increment in productivity that adds to worldwide food, feed, and fiber safety.
- Autonomy with respect to a country's farming land.
- Conservation of biodiversity by preventing deforestation and protecting biodiverse sanctuaries.
- Overcoming tasks that are related to overall climate change.
- Improvement in health and economic status.

2 Worldwide Expanse of GM Crops

In 2016, the accrued hectare (planted since 1996) flowed was documented for 2.1 billion hectors and from 26 countries 7 industrial countries and 19 developing countries were planting GM crops (Fig. 1). It would be noticed that around 185.1 million hectares of GM crops are proportionate to around 20% of aggregate land zone of China with 956 million hectors of US 937 million hectares, and it is more than 7 times the land territory of the UK with 24.4 million hectares [5]. The expansion

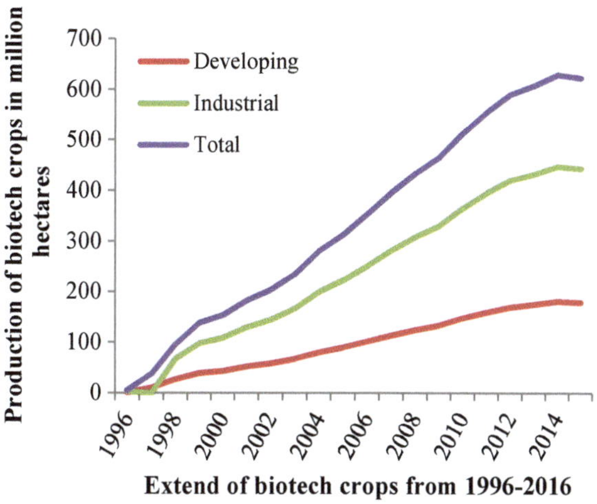

Fig. 1 Worldwide area of biotech crop from 1996 to 2016 (in million hectares): developing and industrial countries (ISAAA 2016)

in 2015 and 2016 is recorded around 3% in equivalent to 13.3 million acres or 5.4 million hectares of land (Fig. 1).

3 Dissemination of Biotech Crops

According to 2016 data, a total of 26 countries are planting biotech crops, and among them 19 are emerging and 7 are from developed countries. All top ten developed countries which are cultivating over 1 million hectares are led globally by USA (39%) while in 2015 around 72.9 million hectares which followed by Argentina (13% around 23 million hectares), Brazil (27% around 49.1 million hectares), Canada (6% around 11.6 million hectares), China (2% around 2.8 million hectares), India (6% around 10.8 million hectares), Pakistan (2% around 2.9 million hectares), Paraguay (2% around 3.6 million hectares), South Africa (1% around 2.7 million hectares) and Uruguay (1% around 0.3 million hectares). Furthermore, 16 countries grow an aggregate of roughly 4.9 million hectares (Fig. 2).

It was testified that eight out of top ten countries (Argentina, Brazil, China, India, Pakistan, Paraguay, South Africa, and Uruguay) are growing about 1.0 million hectares of biotech crops. While in 2016, Burkina Faso and Romania did not plant biotech crops due to modification in cotton germplasm so arduous reporting requirement of biotech crops planting, respectively.

There are totally 18 biotech mega-countries. These are those countries which grow 50,000 hectares or more biotech crops. Among these 18 mega-countries, 14 are emerging countries from Africa, Asia, and Latin America. In 2016, the high extent of biotech mega-countries (18 out of 26 which come to 69%) mirrors the significant deepening, broadening, and stabilizing in GM crops adoption. It has occurred within

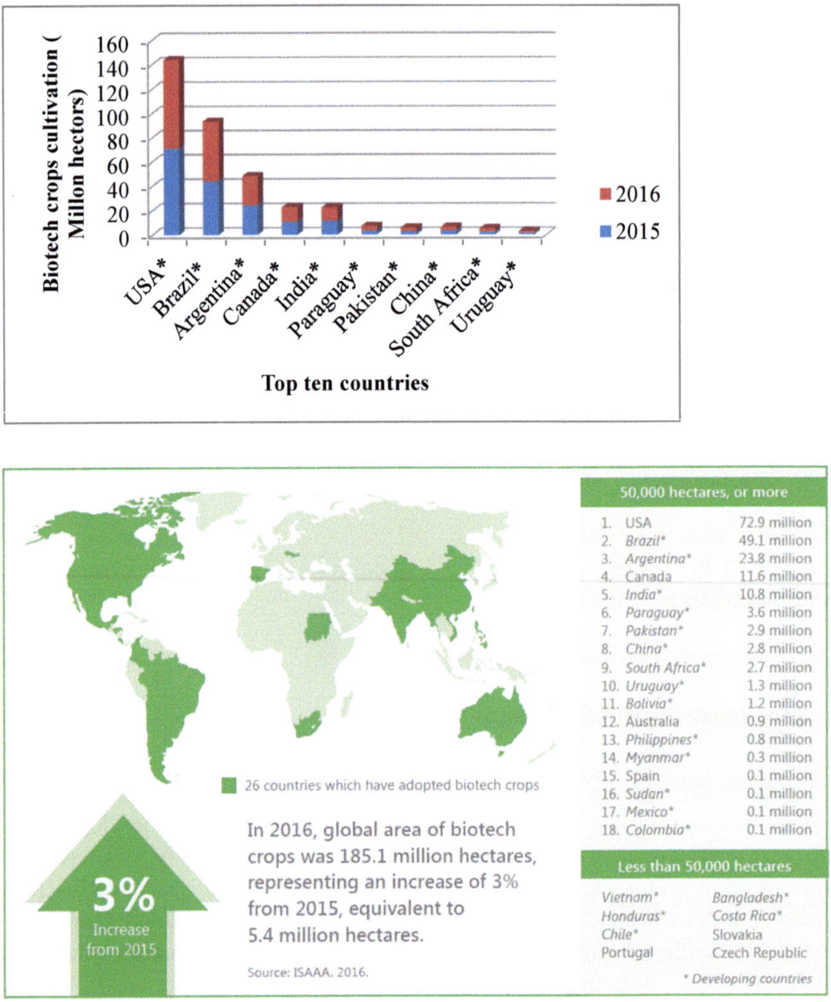

Fig. 2 Worldwide expanse of GM crops (million hectares) from 1996 to 2016, by country, mega-countries, and for the top ten countries (*Source* ISAAA 2016)

the group of more advanced mega-countries that adopted more than 50,000 hectares crops in all six continents [6].

In total hectares, Brazil had the major year-over-year growth with 4.9 million hectares, trailed by USA with 2 million hectares, while Canada with 600,000 hectares, South Africa with 400,000 hectares, and Australia with 200,000 hectares. In terms of worldwide share, million hectares of biotech crops are planted globally, and top three biotech countries are USA (39%), Brazil (27%), and Argentina (13%) for a total of 78%. In 2016, out of 26 countries that planted biotech crops are America (12 countries), Asia (8 countries), Europe (4 countries), and Africa (2 countries). On the basis of hectare for biotech, plantation is 88% from America, 10% from Asia, 2% from Africa, and less than 1% from Europe [6, 7].

Currently, there are ten countries in Latin America, which was showing an advantage from the wide acceptance of GM crops. Countries in descending order of hectare are Brazil, Argentina, Paraguay, Uruguay, Bolivia, Mexico, Colombia, Honduras, Chile, and Costa Rica. While in the Pacific and Asia, there are eight countries planting biotech crops led by India, Pakistan, China, Australia, Philippines, Myanmar, Vietnam, and Bangladesh. From last 6 years, Japan also commercializing biotech flower (commonly known by blue rose) by growing them in green house through more advance techniques then other countries.

Four European countries Spain, Czech, Portugal, and Slovakia are continuing the growth of biotech plants as it led by 17% (136,363 ha in 2016 as compared to 116,870 in 2015) by Spain, Portugal, Slovakia, and Czech Republic. In 2016, Romania decided not to plant biotech crops because of the onerous requirement by the government [8].

4 Monetary Benefits of GM Crops

In the past 20 years (1996–2015) of commercialization, it is noteworthy that other than 167.8 billion US$ farmer gain additional income generated by biotech crops, around 81.7 billion US$ was generated in industrial countries and 86.1 billion US$ in developing countries. In 2015, developing countries had a lower share with 48.7%, equivalent to US$ 7.5 billion of the total US$ 15.4 billion grains with industrial countries at US$ 17.9 billion [7]. From last 20 years of biotech crops commercialization, a total six principal countries have gained economically. In descending order of magnitude are USA (73 billion US$), Argentina (21.1 billion US$), India (19.6 billion US$), China (18.6 billion US$), Brazil (16.4 billion US$), Canada (US$ 7.3 billion), and others (11.8 billion US$) for a total of 167.8 billion US$. In 2015 worldwide economic benefits were 15.4 billion US$. From this total amount, 7.5 billion US$ for emerging and 7.9 billion US$ was for industrial countries (Fig. 7) [7, 8].

5 GM Crop Approvals by Different Countries

GM crops achives the endorsements at the crest in 2014 by issuing the endorsement from 22 countries (Fig. 3). In 2016, a couple of endorsements were allowed to some degree because of different reasons, incorporating changes in country regulations, for example, in the Mexico and Philippines, climate-related issues influencing trial, and the attention on new single occasion as in USA. The USA had just 18 endorsements in 2016, which is the most minimal since 2012.

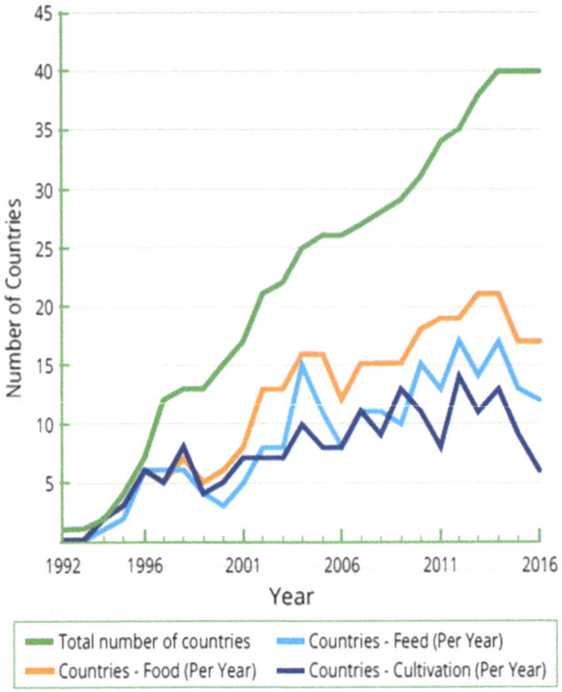

Fig. 3 Countries that issued approvals, 1992–2016 (ISAAA 2016)

6 Number of GM Events Approved

In 2015, high numbers of GM events are permitted with just a couple of occasion endorsements in 2016 (Fig. 4). As this could be a normal eventual outcome of a large number of occasion endorsements in the previous year, but technology developers observed endorsements and adoption in different countries. This could also be a result of the relatively slow release of new GM crops as technology developers put more

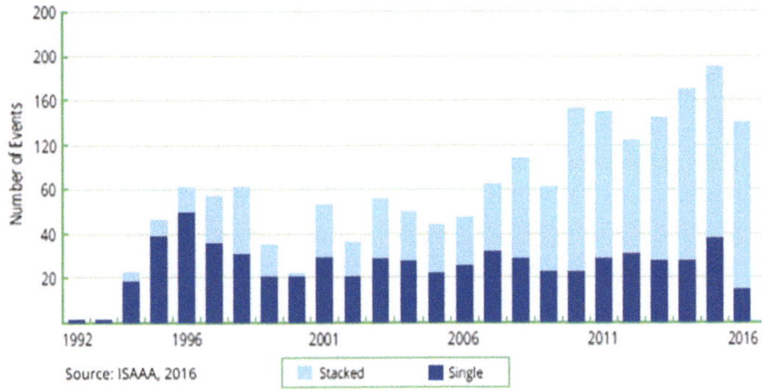

Fig. 4 Number of proceedings sanctioned events, 1992–2016 (*Source* ISAAA)

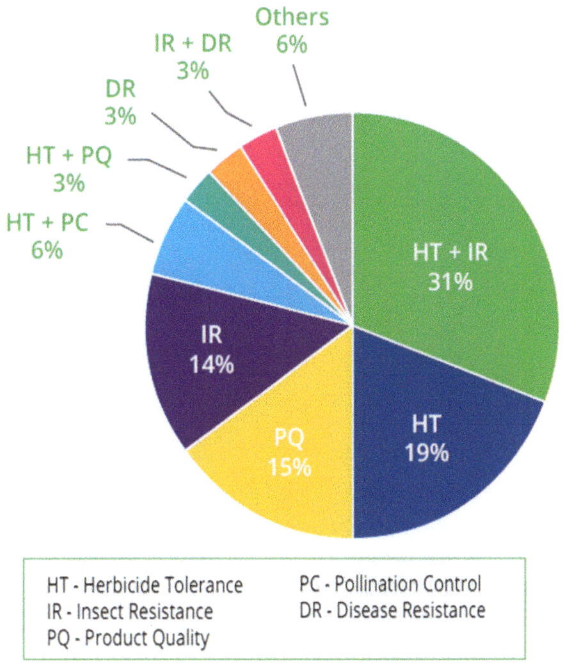

Fig. 5 Trait dispersal in the permitted year, 1992–2016

focus on using new approaches and reproducing techniques in developing improved crops.

In 2016, the greater part of affirmed approved events was stacked or pyramided (Fig. 4). This pattern of stacked events outnumbering the single events started in 2008 and peaked in 2016. This means agriculturists are more selective and choose biotech events/varieties with more traits to offer for cost reduction and better monetary benefit. The same is shown in Fig. 5, in which events with both herbicide tolerance and insect resistance included over 25% of the occasions affirmed, while events with more than one trait made up at least 43% (HT + PC, HT + PQ, and IR + DR) of the endorsed events. This pattern will probably proceed into the future since farmer demands more traits in an event, particularly in maize. It was reported that maize still has the most number of event occasions in the catalogue. This is likely because of the quantity of single maize events, which can be joined with different proceedings to shape the desired occasion.

For 2016, the cumulative nourishment, feed, and cultivation endorsements were 115, 87, and 49, individually (with an overall total of 251). These endorsements are partitioned among 87 events from seven crops and were granted by 16 countries. Among all Argentina had the most astounding number of approvals which were generally from maize events.

In 2016, total feed, food, and agronomy sanctions were 115, 87, and 49, respectively (with an overall total of 251), in 2016. These agreements are divided among 87 proceedings for seven crops and were contracted by 16 countries (Fig. 6). The num-

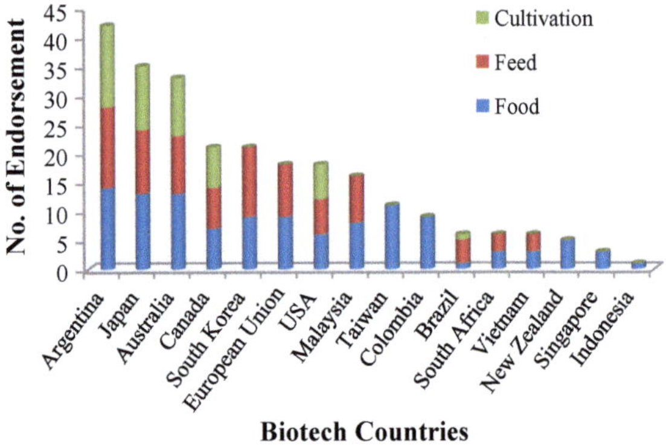

Fig. 6 Endorsement per country for 2016 (*Source* ISAAA GM sanction catalogue)

ber of approvals for 2016 is lower than the previous 2 years (318 in 2015 and 302 in 2014). Argentina had the maximum sum of endorsements which were typically from maize proceedings. Since 2007, the number of stacked proceedings endorsed has dominated the single event approved. In 2016, the stacked proceedings sanctioned are made up of 82.6% of the combined sanctioned proceedings [9].

This is suggestive of the expanding request by agrarians for the event with more characters to additionally build their benefit. From 1992 to 2016, 40 countries have given 1777 food approvals, 1238 feed approvals, and 813 agronomy endorsements, scattered among 477 events from 29 crops (ISAAA GM Approval Database).

7 Imminent of Biotech Crops

The upgrading and release of GM crops follow a borderless trend in terms of crop/trait. This pattern might be considered as a technique by farmers or food developers to gain acceptance of products. The original biotech crops are supplied to farmers to increase yield and resist biotic stresses. The second-age GM crops incorporate stacking IR/HT traits and these traits that can help to mitigate the effects of climate change. While the third era of GM crops incorporate all specifications of both era and includes the supply of Biotech crops to the consumers and manufacturing industry [10].

8 First-Generation Biotech Crops with Agronomic Traits

The first generation of biotech crops primarily involved the exertion qualities, such as herbicide tolerance, insect resistance, and disease. These were positioned in the four noteworthy biotech harvests such as canola, cotton, maize, and soybean. Since 1996,

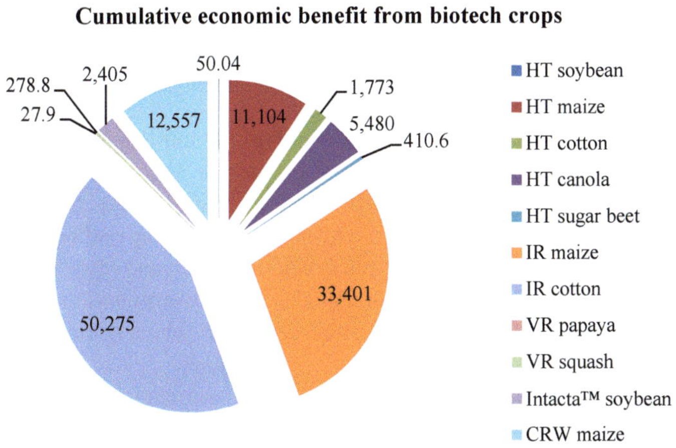

Fig. 7 Economic benefits by trait/crops (million US$), 2015*. *Source* *Brookes and Barfoot 2017 Forthcoming. **ISAAA GM Sanction Catalogue 2016

there has been a various number of single and stacked same trait events (HT/HT and IR/IR) that were popularized with IR cotton and HT soybean, and both crops have the most numbers of events [10, 11].

All these insect resistance crops established a tremendous monetary advantage to ranchers for around 50.3 billion US$ (Fig. 7). IR (Bt) cotton benefits ranchers in different countries such as USA, India, Pakistan, Brazil, and China. The next crop/trait with the highest economic benefits is herbicide-tolerant soybean at 50.04 billion US$. Large hectarages of HT soybean were planted in Brazil, USA, Argentina, Paraguay, and Canada, the top five countries [11].

9 Second-Generation Biotech Crops with Stacked Input Traits

The second-age input traits incorporate different heaps of HT/IR in maize, cotton, and soybean which result in reduced production in expenses and ease in farming. These attributes were released toward the end of the first span (1996–2005) and then the start of the second decade (2006–2015). Advantages for stacked maize were US$ 12.5 billion and the stacked soybean US$ 2.4 billion. Second-age crops also possess traits to address issues related to environmental change, such as cold tolerances, salt interruption, and drought. Drought tolerant in maize, propelled in 2013 in the USA at 50,000 hectares and expanded to ~1.2 million hectares in 2016, can diminish transpiration by 175% under stress conditions.

10 Third-Generation Biotech Crops for Nutrition and Product Advantage

In 2016, world food prize featured about the human nourishment and prosperity, which turned into the basic segment of food production strategy. World Food Prize victors—Dr. Maria Andrade, Dr. Howarth Bouis, Dr. Jan Low and Dr. Robert Mwanga—were granted for their endeavors to elevate the well-being and prosperity of the worldwide poor and malnourished populace. Andrade, Low, and Mwanga of the International Potato Center (CIP) were documented for their determinations in developing "the single most successful example of biofortification" which is the orange-fleshed sweet potato.

Bouis was praised for making HarvestPlus, an association that spotlights on enhancing nourishment and general well-being through biofortification. Similarly, third-age biotech crops are equipped toward enhancing wholesome quality. The deliberation for third-age biotech crops is on creating yield attributes for enhanced item quality and composition, for example, modified/improved oils (omega-3 unsaturated fats and high oleic acid in soybean), transformed starch/sugar (potato), low lignin (alfalfa), non-browning fruits (potatoes, apples) that are already accessible in the market, and expanded betacarotene, ferritin, and vitamin E in real staple harvests, which are in the advanced phases of advancement [12].

Soybean has the most production qualities to suggest clients with a high oleic acid event that reduce trans fats, low phytate. It also improves mineral absorption by the human body and reduces phosphorous levels in animal manure, while high stearic acid and omega-3 reduce harmful fats in humans.

Recently, biotech crops are adopted by farmers so that it can help to reduce food waste and environmental impact. According to FAO data globally enough food was produced to feed everyone on the planet, but virtually 800 million people around the world suffer from hunger because roughly 1.3 trillion tones of food is wasted per year [12, 13].

As per report Environmental Protection Agency (EPA) of US estimates 133 billion pounds of the US food supply (approximately 31%) is wasted annually, and contributing 18% of the total methane emission that comes from landfills. In addition, about half or more than that all vegetables and fruits are wasted each year because consumers prefer crops that always look fresh and unblemished. Therefore, through biotechnology crops are designed with non-browning and non-bruising traits so that it becomes available and greatly eliminate losses due to wastage [13].

11 New Developments and Products in the Pipeline

According to 2016 data, the numbers of GM events for commercial cultivation, regulatory stages get doubled from 2008 to 2014 [9]. The program of new endorsements recorded in the ISAAA GM Approval Database demonstrates the transcendence of insect resistance and herbicide-tolerant traits, and gene stacking in the four major crops.

Crop Life International published a list in 2016 about the stacking of IR/HT traits and a few products of soybean and canola. The published biotech crops, which are in pipeline, are in public sector research institute and are geared toward yield and biomass improvement, disease resistance, improved nutrition and product quality. This indicates that both industry and public technology developers are addressing the needs of both consumers and farmers.

12 Food Safety Regulation

From the last two decades of biotech crops, safety and its commercialization is been impeccable with consumer's health, poultry and livestock, and non-targeted organism. The technical advice to Codex Alimentarius Commission of several experts consultant on the evaluations of GM crops (by The Food and Agriculture Organization (FAO) and World Health Organization (WHO)) so that it inducted in the Codex Guidelines on safety assessment of GM foods.

The safety of GM crops is under WHO and to view public health protection, in close partnership with the Food and Agriculture Organization along with other international bodies. Even though all the records are maintained for biotech by the safety point of view, still some government's bodies and public are still showing hesitation to accept biotechnology. There are reasons other than food safety, which effect public attitudes toward GM foods, including customer decision, control of the worldwide seed market and natural pecking order, and effects on small agriculturists. Such concerns influence the acknowledgment and take-up of the innovation and ought to be recognized, comprehended, and considered amid exchanges about GM items and the advancement of policy [13].

13 Regulatory Barriers to GM Crops Adoption

13.1 Budget and Erection of Regulation

The selection of biotech crops globally depends upon the cost and structure of National Regulatory Organization, but the regulatory process is very expensive. So McDougall revealed that from 2008 to 2012 the expense is about 136 million US$ for the approval of the innovation, improvement, and biotech determined harvest quality [14]. Therefore these timelines and cost become a huge burden to small and medium innovation engineers, and just a couple of vast corporate organizations can continue utilizing the innovation and commercializing the item. Consequently, these turn into an illusory and erroneous view of a native connection among GM and the extensive organizations [14, 15].

13.2 Rigorous and Incompetent Parameter

Efficiency of the regulatory process is vital for the incessant adoption of biotech crops that will profit the entire stakeholder. Both open (local farmers) and private area (Industrial) developers depend on a productive, expectable, and arduous however non- arduous science-based monitoring process as a critical part in the development strategy. The innovative work results contribute to the solutions that address food and agriculture challenges, but product commercialization delay resulted in the hindrance to this solution to the farmers and consumers. Therefore, it effects the growth and progress of farmers and consumers, food manufacture industries, locally and globally. Finally, it will create vagueness among public and private investors in agriculture research and development, or said that the further investment in agriculture innovation is jeopardized.

14 Conclusion

The year 2016 has been surprising for biotech yields with respect to the first runtime because some Nobel Laureates released an announcement in favor of biotechnology along with the critics as precarious deportment against the Golden Rice. Furthermore, according to the agenda of sustainable agriculture vision 2030 aimed to resolve the hunger problem with the help of agencies or governing bodies like UN FAO, IFPRI, the G20 countries, and other similar bodies. Along this several academies likes National Academic of Sciences, Engineering and Medicine (US) are investing several studies on biotech crops and found that the GM crops and conventionally breed crops are similar in respect to possible risk to human health and environment.

Today, biotech crops have a perfect record of safe use and its utilization, while future generations can gain profit by an extensive variety of biotech crops with an enhanced attribute for high yields and nutrition so that all GM crops are safe for utilization and environment. In 2016, the third era of GM crop commercialization was high as 26 countries cultivated around 185.1 million hectors of biotech crop with agronomic traits in 4 biotech crops such as soybean, maize, cotton, and canola. From recent reports, biotech crops are playing a significant role in building up the profitability percentage; hence, it can be served as an engine of economic growth, particularly for the poor farmers. The strategy of sustainable biotech harvests is supported by the numerous Academics of Science around the world, and it permits the expedition of worldwide crop production. Biotech crops are necessary, but it is not a solution of all crop-related problems. Biotech crop needs are same as convention crop needs such as the practice of rotations and resistance management for insects, pathogens, and weeds. Moreover, biotech crops are profitable but still have some challenges which are incorporated as:

1. The governing barrier limits the scientific innovation and sometimes restricts the technology development that may benefit the farmers and consumers.

2. Distribution in the growth trade as it brought by synchronous approvals and thresholds on the low-level presence in GM crop-trading countries.
3. The requirement for continuous discussions among the stakeholders for quick understanding and raise of biotechnology emphasizing stressing benefit and safety. So, the better way to communication is the use of social media. To overcome these challenges is a difficult task that requires the collaborations among government and private sectors. Collaboration or partnership could be an assurance for the sufficient and nutritious supply of food for everyone.

Acknowledgments The authors acknowledge the gracious funding provided by the postdoctorate fellowship by the University Grant Commission, Government of India. The authors are also thankful to Dr. Ajay Mathur, Director General, TERI, for his kind support.

Conflicts of Interest The authors declare no conflict of interest. The founding sponsors had no role in the design of the study; in the collection, analyses, or interpretation of data; in the writing of the manuscript; and in the decision to publish the results.

References

1. Bloomberg Markets: Americans are buying gene-edited food that's not labeled GMO. https://www.bloomberg.com/news/articles/2016–07-14/gene-edited-canola-oilarrives-without-gmo-style-shopper-outrage (2016)
2. Business Wire: Global genetically modified seeds market to witness growth through 2020 due to rise in adoption of bio-fuels: Reports Technovio. http://www.businesswirecom/news/home/20160830005089/en/Global-Genetically-Modified-Seeds-Market-Witness-Growth 30 August 2016
3. Crop Biotech Update (CBU): http://www.isaaa.org/kc/cropbiotechupdate/default.asp
4. GMO Answers: How GMOs help us reduce food waste and its environmental impact. http://www.forbes.com/sites/gmoanswers/2016/11/18/gmos-help-reduce-food-waste/#37e8b13d6546 (2016)
5. McDougall, P.: The cost and time involved in the discovery, development and authorisation of a new plant biotechnology derived trait. https://croplife.org/wp-content/uploads/2014/04/Getting-a-Biotech-Crop-to-Market-Phillips-McDougall-Study.pdf (2011)
6. MOEF&CC: Minutes of GEAC meeting 126, 127, 128, 129, 130 and 131, the Genetic Engineering Appraisal Committee, the Ministry of Environment, Forests and Climate Change (2016)
7. Brookers, G., Barfoot P.: GM crops: global socio-economics and environmental impacts (1996–2015) (2017)
8. NBC News. 17 May 2016: Genetically modified crops are safe, report says. http://www.nbcnews.com/health/health-news/genetically-modified-cropsare-safe-report-says-n575436
9. Parisi, C., Tillie, P., Rodríguez-Cerezo, E.: The global pipeline of GM crops out to 2020. Nat. Biotechnol. **34**, 31–36 (2016). https://doi.org/10.1038/nbt.3449
10. Phillips PWB: Economic consequences of regulations of GM Crops. https://www.geneticliteracyproject.org/2014/12/11/economicconsequences-of-regulations-of-gm-crops/ (2014)
11. UNEP GEF: National Biosafety Framework Myanmar, Ministry of Agriculture and Irrigation, Department of Agricultural Planning, Development of Biosafety Framework Project Myanmar (Nov 2006)
12. USDA FAS GAIN Reports for 2016. https://gain.fas.usda.gov/Recent%20GAIN%20Publications/Forms/AllItems.aspx

13. Butler, T., Reichhardt, T.: Long-term effect of GM crops serves up food for thought. Nature **398**(6729), 651–653 (1999). https://doi.org/10.1038/19348
14. Symth, S.J.: The costs of regulatory delays for genetically modified crops. J. Int. Law and Trade Policy. **1**(7), 173–195 (2006)
15. Gardner, J.G., Nehring, R.F., Nelson, C.H.: Genetically modified crops and household labor savings in US crop production. AgBioForum **12**(3&4), 303–312 (2009)

Global Food Security: A Truncated Yield of Underutilized and Orphan Crops

Anoop Anand Malik and Gulshan Chaudhary

Abstract Orphan crops and underutilized crops are considered as the major staple food crops in many developing countries because of their particular role in food security, nutrition, and foremost income generation to resource-poor farmers and consumers. Alike other crops, orphan crops are also categorized under cereals, legumes, root crops, and fruit crops. Orphan crops are in general more adapted and acclimatized to the innumerable abiotic stress than the major crops of the world. However, due to lack of exposure to new era of biotechnological practices and techniques, orphan crops produce inferior qualitative and quantitative yields. According to crop science society of America data indicated 12,650 edible plant species exist and only 30 of them are being utilized by 95% population of world. These inclined selections of limited crop production will lead to the breakdown of United Nations goal of ending global hunger by 2030. Biotechnology interventions and its new era expansions in the field like tissue culture, marker-assisted breeding, in vitro gene modulations etc. are some important capacities to upsurge the productivity of planting material and, hence, could eventually increase the global consumption. These techniques have contributed tremendously to the safeguarding, improvement, and distribution of orphan crops, especially the vegetatively propagated crops. Market surveys revealed these orphan crops are an important source of household incomes and substantially contribute to global poverty reduction. The robust research is required to fill knowledge gaps and snags for most of the identified species as no concrete scientific data is globally available.

Keywords Food security · Orphan crops · Biotechnology · Underutilized crops
Nutritional security · Marker-assisted selection · DNA markers
Molecular genetics · Agricultural policies

Anoop Anand Malik and Gulshan Chaudhary contributed equally to the chapter.

A. A. Malik
TERI School of Advanced Studies, New Delhi, India
e-mail: anoopanandmalik@gmail.com

G. Chaudhary (✉)
The Energy and Resources Institute, New Delhi, India
e-mail: chaudhary84@gmail.com

M. Khoobchandani and A. Saxena (eds.), *Biotechnology Products in Everyday Life*, EcoProduction, https://doi.org/10.1007/978-3-319-92399-4_11

1 Introduction to Orphan and Underutilized Crops

Orphan crops are the least investment minor crops and neglected species in the field of agricultural science which have a great prospective for food production and development worldwide. The inadequate appreciation of their potential and incessant exploitation could be one of the major issues for its low agricultural value [1, 2]. They should be better described as "minor crops" for the reason that there low commercial value of production and trade as compared to the major crops in the world. In addition, these orphan and underutilized crops receive low attention toward the research and development, therefore, not enough scientific information is available in the public domain. Hence, these yields remain underutilized for further improvements due to ignorance as neither the private and nor the public sector has invested significantly [3]. An Important key points to insistence about orphan crops are that, they vary in different geographical vicinities. There are few significant crops, which could be of high importance and play a major role for global advancement. The less usage and limited markets could be the real reasons for our focus which can be legitimized to advance the generation and usage of these minor yields as portrayed beneath [4].

- It represents colossal revenue of agro-biodiversity.
- It would be of great potential to improved incomes, food security, and nutrition.
- It can also solve the problem of "hunger in the world."
- It has no authentication for seed supply.
- It gets considerably less attention from research, extension services, farmers, policy and decision makers, donors, technology providers, and consumers.
- Most of them are highly nutritious and/or have medicinal properties or other multiple uses.

Furthermore, India is a developing country and agriculture is considered to be a backbone of Indian economy. This chapter is focused on the challenges and opportunities for using several orphan harvests in India and worldwide. The total area harvested and total production of orphan crops were observed from Food and Agriculture Organization (FAO) of the USA database for two different sets of countries and India [5]. The country dataset comprises 15 developed countries, 45 developing countries, and India to compare the yield status (Fig. 1).

Orphan and underutilized crops in India are diverse because of their genetic, agro-climatic, and economic niches. These viewpoints gave rise to a new series of questions regarding the orphan crops. All biotechnological interventions and research on major species nowadays being used as research models—such as wheat, rice, or Arabidopsis—play an important role for forthcoming improvements of orphan crops and help in a hike of agricultural economy. And slowly but surely, it will be possible to compare major crop benefits over the orphan crops in the near future in India.

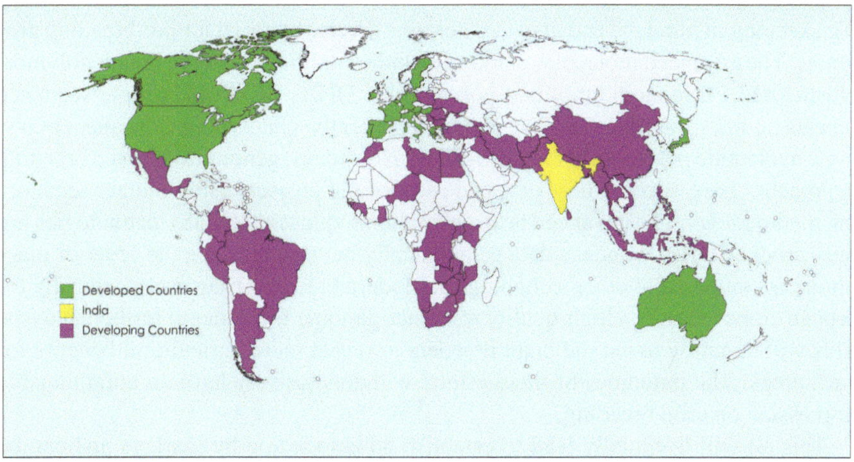

Fig. 1 Detailed world map showing the two major groups of developed and developing countries along with India as the source of data collection for area harvested and total orphan crop production (*Source* FAO database)

2 Conceivable Biotechnological Interventions for Boosting Orphan and Underutilized Harvests

Agricultural biotechnology is generating new implements to wrestle with the issue of rustic destitution, employment, and income generation by upgrading the farm productivity and production, enhance quality, and investigate showcasing openings in more up-to-date ways. Among other fields of agriculture science, biotechnology is a rapidly developing field, which could be helpful in an attempt to meet the current and emerging challenges, for example, poor sustenance, unstable and constrained food production, and confined fuel availability. Indeed, there are few sectors, which cover the foremost part of agricultural biotechnology.

(a) Molecular marker characterization of genetic diversity
(b) Tissue culture and micro-propagation
(c) Marker-assisted selection (MAS)
(d) Genomics and the related disciplines of proteomics and metabolomics
(e) The production of transgenic crops [6].

The description of few advanced techniques including the use of micro-propagation for in vitro vegetative multiplication through somatic embryogenesis or organogenesis to clone huge numbers of plants from genotypes of predominantly desirable traits, should be used more broadly. The upcoming technologies and developments in the field of genetics and genomics provide a more unified understanding of the biology of plants, which as a result can further emerge new opportunities for applying advanced science to orphan crops. Moreover, portrayal of genetic diversity by molecular markers is critical for formulating effective sampling strategies, e.g., keeping in mind the end goal to determine diverse material for pre-breeding programs. The advanced molecular markers include amplified fragment length polymorphism (AFLP), random amplified polymorphic DNAs (RAPDs), simple sequence repeats, or microsatellites (SSRs); and, more recently, genotyping by sequencing–single nucleotide polymorphism (GBS-SNP). Nowadays, genomics-assisted breeding approaches have successfully swapped most of the projects using Sanger sequencing methodology. The advanced next-generation sequencing (NGS) technologies not only produce DNA sequence data economically but also enhance the order of magnitude as compared to other technologies. Modern breeding program specifically for orphan crops requires a high-quality reference genome for in-depth further analysis. This will certainly assist the plant breeders to reveal more agricultural benefits for such crops. The outcomes of these efforts will conclusively have an unfathomable impression on crop breeding.

This all will eventually lead to establish an association by markers and can be used to understand complex traits, which further assist in selection for traditional breeding methods. The result of genetic engineering, a transgenic, is one that changed by the addition of at least one transgene from another, often unrelated organism. These foreign genes add to a scope of properties including resistance/tolerance to biotic and abiotic factors and enhance the nutritional status of the crops. Genetically modified/transgenic/engineered plants can be used as sources of new germplasm or cultivars, which acts as a new source of variation in breeding programs and extremely useful as proof-of-concept tools to dissect and characterize the activity and interplay of gene networks for biotic/abiotic stress resistance [7]. The method to combine a high-throughput whole-genome re-sequencing with a bulked-segregant analysis represents a quicker and more efficient method of identifying a target gene. Therefore, the functional orthologs of gene with mutations could also be selected so-called as “green revolution genes” in rice and barley [8]. In spite of significant talk with respect to the potential employments of biotechnology for addressing worldwide agrarian difficulties, practical deployment for a vagrant harvest is currently limited [9]. The biotechnological interventions are required to advance and inflate the formation of nourishment products to meet the dietary requirements of the families. The objective of strategy system is to decrease neediness, sustenance security, social insurance change, industrialization, and assurance of the earth through the protected utilization of biotechnology. The real concentration range for the approach is the improvement of the preservation and use of biodiversity. The strategy for attaining the goals is the use of biotechnological tools to characterize primitive plants and to assess their economic potentials for biotechnological applications as it explains the non-competitiveness and underutilization of orphan crops; also, it has insufficient scientific evidence about their potential value. Other than this, now government is additionally offering an opportunity to the utilization of present-day advanced biotechnology strategies for the change of orphan crops and its preservation for their hereditary assets as its more productive route for the choice of increases for phenotypic portrayal.

The objective of biotechnology is to comprehend the genetic and molecular origin of every biological process in orphan and major crops that are pertinent to the species. Furthermore, it is premier requirement to develop a set of indicators, and a system for observing biotechnology interventions, by which the costs and benefits of activities can be fully assessed, especially costs and benefits for smallholder producers [10].

3 Interest in Logical Research for Development for Orphan Crops

The vagrant harvests advancement is needed to build interest in logical research and development since there is considerably less attention from international or regional crop research organizations (such as CGIAR, the Consultative Group on International Agricultural Research). To meet the requirement of hunger and the global development, orphan crops demand for substantial investments in scientific research and innovations. The multilateral treaties and agreements like Convention on Biological Diversity (CBD) and Cartagena Protocol on Biosafety respectively ensures to conserve biological diversity, to use biological resources sustainably, and to share equitably the benefits arising from the use of genetic resources. But then again in case of India, government should pay more attention over this issue by a sort of unique program to concentrate on logical research on yield-related traits and advancement for all states orphan crops either by traditional or advanced GM technologies for the betterment of food security level in the near future [11].

4 Change in the General Population Expansion Framework

According to the present scenario, the population growth throughout the following 30 years will be gathered solely in the developing countries, where more than 1 billion individuals as of now live on less than US$ 1 per day, more than 800 million individuals are underfed, and 200 million children are underweight [12]. The worst condition was observed in rural areas where agriculture was the prominent foundation of earnings and payments. Due to extremely low agricultural investments by the public and private divisions, there were many poorest and neglected regions in the world in crop production. There is a strong need to remove the bleak from some societies by introducing certain mechanisms which can help in fueling the advanced agriculture practices. The latest uploaded data (2014) for orphan crops on FAO was observed to compare the total area harvested (ha), yield (ton/ha), and production (tonnes) between developed, developing countries (like China, Afghanistan, Sri Lanka, Kenya, South Africa, Iraq, Bangladesh, Pakistan, Turkey, Georgia, Zambia) and India [5] (Table 1).

The data observed was directing toward the fact that the yield and area harvested on production of orphan crops were comparatively low in developing countries which could be a major problem in the near future in nutrition deficiency, loss of biodiversity, food security, etc. The precise and latest data was available for only seven orphan

Table 1 Area harvested, yield, and production values of few orphan crops

Crop	Developed countries[a]			Developing countries[b]			India		
	Area harvested (ha)	Yield (ton/ha)	Production (tonnes)	Area harvested (ha)	Yield (ton/ha)	Production (tonnes)	Area harvested (ha)	Yield (ton/ha)	Production (tonnes)
Beans, dry	59,802	2.190	118,871.47	283,059.18	1.399	317,071.93	10,000,000	0.453	4,110,000
Buckwheat	11,258.93	0.543	15,088.73	21,550.18	0.224	19,884.80	0	0	0
Chickpeas	48,076.13	0.795	62,335.53	60,199.38	0.675	46,152.84	9,927,000	1.097	9,880,000
Fonio	0	0	0	10,466.27	0.055	10,786.09	0	0	0
Millets	15,587	1.143	26,673	237,155.84	0.989	172,774.04	8,904,000	1.414	11,420,000
Mustard seed	15,112	0.517	15,936	8096.98	0.206	6866.53	0	0	0
Pigeon peas	0	0	0	25,132.71	0.320	29,033.51	5,602,000	0.647	3,290,000

Source of entire table FAOSTAT (2014)

[a]A total of 15 developed countries were categorized for the analysis

[b]A total of 45 developing countries excluding India were categorized for the analysis

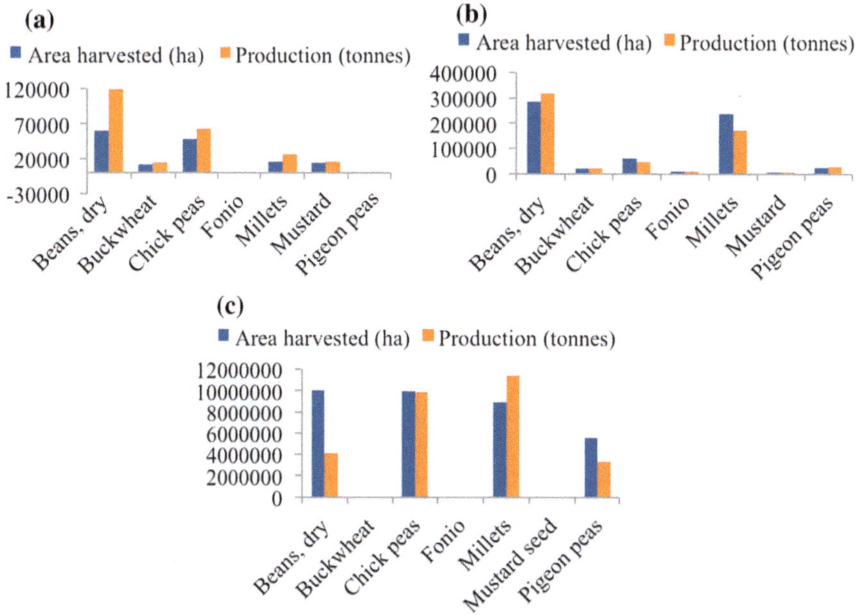

Fig. 2 Graph plotted with the values of total area harvested and total production of orphan crops for **a** developed countries, **b** developing countries, and **c** India

crops namely beans (dry), buckwheat, chickpeas, fonio, millets, mustard seed, and pigeon peas (Fig. 2).

The values in the table were further analyzed by calculating the average values in each set of developed and developing countries. The unavailable data in a particular country was marked as zero (0) in the table for average estimations. For the imperative part that vagrant products play in agricultural advancement, respected authorities in the government ought to keep up and educate people in general horticultural augmentation framework. This will further lead to address the necessities for poor people in the country who have sustained and keep on growing these yields [13].

5 Conservation and Security of Orphan Crops by Reinforcing the Seed Framework

In respect of conservation and security of crops, WFP goals targeted to maintain the genetic diversity of seeds, cultivated plants including through soundly managed and diversified seed and plant banks at the national, regional, and international levels. Furthermore, this helps in promoting the access to fair and equitable sharing of benefits arising from the utilization of genetic resources and associated traditional knowledge at international level [14]. The conservation is managed at two major sectors of the society, the public sector which has a relative advantage in discovery, while the private sector has a comparative lead in overcoming regulatory issues and distributing seeds. Some challenges are that seed for most of these harvests is not readily available on the market to the farmers. Those ranchers who still cultivate these crops rely on farmer saved or recycled seeds, which are frequently of low-grade quality. Though, nearly that is a challenge, it additionally creates an opportunity for developing farmer-driven seed frameworks for these harvests. The farmers who have conserved these crops would be authorized and acknowledged for their contribution [15]. This conservation dexterity needs for the conservation of pure lines of crops for the future research to enhance its nutritional value and other potential factors. Consequently, seed is the most significant contribution to any crop production in an agrarian framework. So seed approach headway is crucial, as seeds are necessities of agriculturists for developing vagrant crops with an untainted assortment. With the goal that the more extensive utilization of vagrant yields is utilized to expand salaries for the provincial poor, the National Export Strategy ought to be returned to intentionally incorporate national vagrant products. Above all else, the system should go for provocative prerequisite and markets for these harvests and their items in the neighboring nations where shoppers have comparable taste. Conservation and security of orphan crop could be pragmatic by strengthening the seed framework in the agricultural production, agro-preparing, and marketing strategy [16].

6 Embrace and Implement of Various Policies on Food and Agriculture

Many studies revealed that the nourishment and farming-related strategies can be utilized to stimulate the use and generation of vagrant harvests in national improvement program. Agrarian policies are measures related to the domestic farm sector and exchange of farming products. Agrarian policies address an extensive variety of issues, including giving adequate sustenance at sensible costs for customers, guaranteeing nourishment security, and enhancing natural quality. Organization for Economic Co-operation and Development (OECD) is to promote policies that will improve the economic and social well-being of people around the world as it provides a forum in which governments can work together to share experiences and seek solutions to common problems. They work with government to understand economic, social, and environmental changes. They measure profitability and worldwide streams of exchange and venture. OECD has created agribusiness bolster pointers that, despite this diversity, express policy measures with numbers in a hasty way to across time and between countries [17].

The International Food Policy Research Institute (IFPRI) teams up with accomplices around the globe; it includes advance implementers, public institutions, private segment, and farmers' organizations to ensure that local, regional, national, and global food policies are based on evidence. IFPRI is a member of the CGIAR Consortium. Recently, FAO Project is generously funded by the Italian Government and executed in close coordination with the Syrian Ministry of Agriculture and Agrarian Reform (MAAR), supporting the establishment of a cadre of professional agricultural policy analysts for the National Agriculture Policy Center (NAPC) and other institutions involved in the Syrian agricultural policy-making process.

According to FAO data in India "The National Agroforestry Policy" was adopted on 2014 under category crop development and diversification, forestry. The policy includes adverse policies, weak markets, and a dearth of institutional finance. Agroforestry has the potential to achieve sustainability in agriculture while optimizing its productivity and mitigating climate change impact. Twelfth Five-Year Plan (2012–2017) emphasized on faster, more inclusive, and sustainable growth. It proposes a two-pronged strategy focusing initially on the need to bring the macroeconomic imbalances under control and to reverse the slowdown, while also pushing for structural reforms in many areas that are critical for maintaining medium-term growth. The Twelfth Plan has set a target of 8% growth over the 5-year period from 2012–13 to 2016–17 [18]. Indian Government launches Pradhan Mantri Fasal Bima and Paramparagat Krishi Vikas Yojana to address the critical importance of soil and water for improving agricultural production.

The policy structure that gives chance to the protection, improvement, and utilization of orphan crops is the National Biotechnology and Biosafety Policy (NBBP). The objective of this structure is to add the national advancement goals such as poverty eradication, improved health care, food security, industrialization, and eco-friendly environment through the use of biosafety markers of biotechnology for conservation of their genetic resources to identify useful variants for conventional plant breeding. The National Environment Management Policy for Uganda (NEMP) is working successfully in Uganda promoting orphan crops in national development, and its aim is to ensure the conservation and sustainable management of Uganda's biological diversity. The other agency like African Orphan Crops Consortium (AOCC) aim is to promote orphan crop cultivation.

7 Conclusion and Future Aspects

As per the updated status available with World Food Program (WFP 2017), women represent 60% of all undernourished people in the world and 70% of the world's poorest people live in rural area and depend on agriculture for their livelihood. The nutrition data also stated that 2 billion people suffer from micronutrient deficiencies and 45% of the deaths among children under age of 5 have malnutrition as an underlying cause from WFP in 2015 which could be a major reason to promote the agriculture practices for orphan crops. It has been noted only 24% of the world's productive lands are degraded for agricultural practices, but this degraded land greatly affects the lives of 1.5 billion people across the globe. From the most recent information of FAO (2012), it was discovered that the populace expanded from 2.5 billion in 2010 to 6.9 billion in 2050. UN populace projection indicates that the world aggregate populace scopes to 9.15 billion of every 2050 so expecting 2.25 billion over next 40 years. This expansion populace in per capita utilization will affect on a world agribusiness by it rate of development as thought about the past. Consumption of food in kcal/person/day is an important key variable used for evaluating and measuring the evolution of the world food situation. The world has gained critical ground in raising sustenance utilization per individual because in last three and a half decades it expanded from normal of 2370 to 2770 kcal/individual/day and the development was joined by huge auxiliary change. This indicates that diets shifted toward more livestock products, vegetable oils, etc., and away from staples such as roots and tubers. Therefore, it can be concluded that the world as a whole the pressures on agriculture to produce more food for the growing population will increase beyond 2050. It comes about that worldwide agrarian creation would need to develop at 0.4% for each year from 2050 to 2080, i.e., not as much as a large portion of the development rate anticipated for the period 2005/2007–2050. To meet the need of global agriculture, it is important to improve the policy perspective and integrate new programs in the legal and safety areas with the help of external aid and consultation to facilitate the adoption of hi-tech agriculture technology. The arrangement needs the foundation of legitimate and security measures as it is more vital that allotting extra subsidizes to biotechnology to improve in crop research by using molecular markers and mapping strategies for a regular harvest rearing, and the involvement of private segment participation useful in distributing seeds. The government goals have been mostly coordinated at advancing the creation and use of significant products which need to be concentrated on the poor people group's yields—the vagrant harvests. In this regard, orphan products can perhaps upgrade the nourishment and sustenance security, as well as empowerments to the farmers. Regardless, if given the required supportive bolster in the regions of agrarian research, augmentation, agro-preparing, and promoting many vagrant products gear up the extraordinary market potential broadly worldwide. This is the ideal time to ensure sustenance and occupation security of our rustic poor by putting resources into their products. On the other hand, in advancing orphan and underutilized crops, it is vital for government to look back their financial aspects. The key role played by these crops in assuring the food and nutrition security of the rural poor people as their role in maintaining and increasing biodiversity which is critical for sustainable environmental management and agricultural production, its adaptability to marginal settings and their ability to stabilize fragile environments are critical considerations in themselves which merit these minor crops for getting more policy support than they are currently being given. More operative plant husbandry and involvement of advanced biotechnological technologies (like marker-assisted breeding technologies, transgenic approaches, high-throughput sequencing) should be implemented to enhance the genetic and genomic resources of vagrant crops. The yield of orphan crop plays a significant role in ensuring sustenance and nourishment security of the country's poor people group. Orphan crops should be as much as biological diverse to fulfill the needs of farming generation as these crops also have the capacity to balance out a delicate situation to ensure nourishment and solve the problems related to food security worldwide.

Acknowledgements The authors acknowledge the gracious funding provided by the postdoctorate fellowship by University Grants Commission, Government of India. The authors are also thankful to Dr. Ajay Mathur, Director General, TERI, for his kind support.

Conflicts of Interest The authors declare no conflict of interest. The founding sponsors had no role in the design of the study; in the collection, analyses, or interpretation of data; in the writing of the manuscript; and in the decision to publish the results.

References

1. Naylor, R.L., Falcon, W.P., Goodman, R.M., Jahn, M.M., Sengooba, T., Tefera, H., Nelson, R.J.: Biotechnology in the developing world: a case for increased investments in orphan crops. Food Policy **29**, 15–44 (2004)
2. Dawson, I.K., Hedley, P.E., Guarino, L., Jaenicke, H.: Does biotechnology have a role in the promotion of underutilised crops? Food Policy **34**, 319–328 (2009)
3. Kamei, C.L.A., Severing, E.I., Dechesne, A., Furrer, H., Dolstra, O., Trindade, L.M.: Orphan crops browser: a bridge between model and orphan crops. Molecular Breeding, **36** (2015)
4. Naluwairo, R.: Investing in orphan crops to improve food and livelihood security of Uganda's rural poor: policy gaps, opportunities and recommendations. ACODE Policy Res. Series **43** (2011)
5. FAOSTAT. FAOSTAT database
6. FAO: Preliminary review of biotechnology in forestry, including genetic modification. Forest Genetic Resources Working Paper FGR/59E (2004)
7. Jewell, M.C., Campbell, B.C., Godwin, I.D.: Transgenic plants for abiotic stress resistance. Springer-Verlag, Berlin Heidelberg (2010)
8. Dong, W., Wu, D., Li, G., Wu, D., Wang, Z.: Next-generation sequencing from bulked segregant analysis identifies a dwarfism gene in watermelon. Scientific Reports 2018, 8, https://doi.org/10.1038/s41598-018-21293-1
9. Jain, S.M., Gupta, S.D.: Biotechnology of neglected and underutilized crops, Springer Science + Business Media Dordrecht (2013)
10. Tumushabe, G.W.: The precautionary principle, biotechnology and environmental litigation: complexities in litigating new and emerging environmental problems. ACODE Policy Res. Series **3** (2001)
11. Naluwairo, R., Tabaro, E.: Promoting food security and sustainable agriculture through facilitated access to plant genetic resources for food and agriculture: understanding the multilateral system of access and benefit sharing. ACODE Policy Res. Series, **18** (2006)
12. Smil, V.: Feeding the world: a challenge to the 21st century. MIT Press, Cambridge (2000)
13. Brookes, G., Barfoot, P.: Global impact of biotech crops: socio-economic and environmental effects 1996–2006. AgBioForum **11**, 21–38 (2008)
14. Goals, G.: The global goals. http://www.globalgoals.org/global-goals/no-hunger/ (2017)
15. Kaushik, S.: 5 schemes introduced by the government to make agriculture a profitable sector. Food and Agriculture 2016
16. Mugyenyi, O., Zeija, F.: The east african customs union protocol: an audit of the stakeholders' participation in the negotiation process. ACODE Policy Res. Series, **14** (2006)
17. Agricultural Policy Monitoring and Evaluation: http://www.oecd.org/tad/agricultural-policies/monitoring-and-evaluation.htm (2017)
18. Twelfth Five Year Plan (2012–2017) Faster, more inclusive and sustainable growth. Planning Commission (Government of India). SAGE Publications India Pvt Ltd. (2013)

Part VI
Ordinary to Extra-ordinary

Metabolomics: An Emerging Technology for Soybean Improvement

Juhi Chaudhary, Rupesh Deshmukh, Zahoor Ahmad Mir and Javaid Akhter Bhat

Abstract Metabolomics is one of the most emerging technologies being used for crop improvement. It allows a comprehensive understanding of complex networks of biological processes, which will certainly help to accelerate the sustainable crop production. In a short time, valuable efforts have been made in developing high-throughput instruments, online tools, and databases for its global application. Soybean is the most widely grown protein/oilseed crop in the world and also used as processed foods, raw material for industrial and pharmaceutical applications and for the production of biodiesel. Due to the high demand for soybean, metabolomics has been applied to understand plant response to different biotic and abiotic stresses and to improve soybean yield. Currently, there is a demand for the development of computational tools and databases for data processing and analysis of metabolic information. Although several large-scale datasets for other 'omics' approaches (such as genomics and transcriptomics) are publicly available, very few studies have been conducted to integrate the genomics–metabolomics approaches to identify QTL/genes. In an effort to comprehend the current scenario and prospects in soybean metabolomics, this chapter highlights the details of soybean metabolomics studies and advances being made to improve environmental stress tolerance. In addition, we discuss pros and cons of different metabolomics approaches. The potential integrated approaches incorporating genomics, transcriptomics, proteomics, ionomics, and metabolomics for soybean improvement were also discussed.

J. Chaudhary (✉)
College of Agricultural and Life Sciences, University of Florida,
Gainesville, FL 32603, USA
e-mail: juhi.chaudhary@gmail.com

R. Deshmukh
National Agri-Food Biotechnology Institute, Sahibzada Ajit Singh Nagar,
Punjab 140306, India

Z. A. Mir
National Research Centre on Plant Biotechnology, PUSA Campus,
New Delhi, India

J. A. Bhat
School of Biotechnology, SKUAST-J, Chatha, Jammu, India

M. Khoobchandani and A. Saxena (eds.), *Biotechnology Products in Everyday Life*,
EcoProduction, https://doi.org/10.1007/978-3-319-92399-4_12

Keywords Metabolomics · Soybean · Environmental stress · Seed composition Omics

1 Introduction

Metabolomics is an important progression to contemporary omics technologies such as genomics, proteomics, transcriptomics, ionomics, and phenomics those are being used for the crop improvement. Some of the advantages of metabolomics include the ability to study the changes in the entire metabolite profile and understand the role and regulation of different metabolites that leads to enhancing yield, abiotic and biotic stress tolerance, disease resistance, seed composition, and flavor enrichment. Furthermore, metabolites are the most predictive of the phenotype because they are the end products of gene expression and protein translation [1]. Plants produce enormous numbers of metabolites that play important roles in plant growth, development, and response to environments. A range of metabolites are responsible for crop yield and quality, also nutrition and energy source for human and livestock [2]. Plant metabolites are broadly categorized as primary and secondary metabolites, the former is responsible for the growth and development of a plant while the latter is essential for plant survival under stress conditions [3]. Metabolite profiling can provide an understanding of the biochemical function of plant metabolism [4]. It involves the measurement of all small molecules, such as sugars, amino acids, and lipids, which are the products and substrates of chemical reactions within biological systems that carry an imprint of all genetic, epigenetic, and environmental factors [5]. Such analyses offer a direct link between a gene sequence and the function of the metabolic network in plants. Hence, metabolite profiling allows interpretation of connections that arise mainly through regulation at the metabolic level [6] (Fig. 1).

Metabolomics progress has benefited from technological advances in mass spectrometry (MS) over the last few years, and the amount and complexity of the

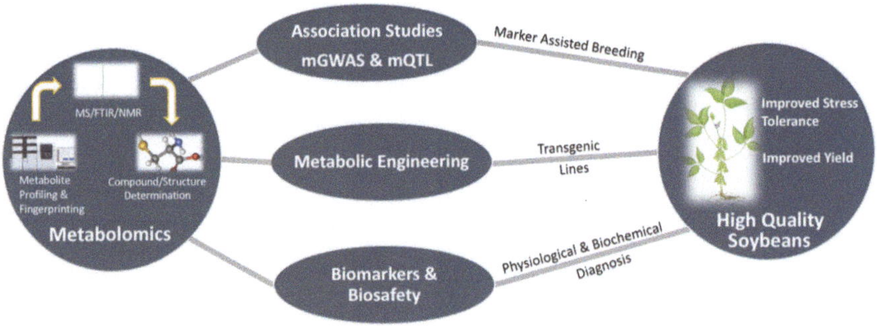

Fig. 1 Schematic representation of metabolomics analysis and its utilization for the soybean improvement

methodological standards produced have been growing rapidly. Several methodological applications have been introduced for metabolite profiling, including Fourier-transform ion cyclotron resonance (FT-ICR-MS), Fourier-transform infrared spectroscopy (FT-IR), gas chromatography (GC-MS), liquid chromatography (LC-MS), capillary electrophoresis mass spectrometry (CE-MS), electrospray ionization (ESI-MS), and nuclear magnetic resonance (NMR) [7]. Despite the advances in modern technologies for metabolome analysis, the major challenge is to develop computational and statistical methods that can decode the information in the large-scale datasets. The computational programs such as MZmine [8], MS-DIAL [9], XCMS [10], OpenMS [11] and other specialized programs have been used as the pipeline of the metabolomics workflow [12]. In addition, there are custom software and Web resources such as Metabolic Networks (MetNet), Plant/Eukaryotic and Microbial Metabolomics Systems Resource (PMR) [13], and Plant Metabolite Annotation Toolbox (PlantMAT) [14] which have been developed to address the biggest challenge, i.e., identification of metabolites. Due to advances in availability of tools and methods for metabolic profiling, metabolic profiling has been employed in many biological studies, for example, metabolite profiling to discriminate specific compounds [15], identification of candidate gene(s) and its function in compound synthesis [16], response in abiotic stress [17], and as well as in the improvement of breeding materials [18].

Despite the above-mentioned advantages, a stand-alone omics technology cannot provide a full understanding of biological processes to improve the environmental stress tolerance. Hence, integration of metabolomics with other omic technologies is an important step for the crop improvement [19]. The accessibility to the 'omics' data has further fueled the progress in plant metabolomics, and efforts have been made to develop tools to integrate genomics, transcriptomics, and metabolomics data, for example, KaPPA-View 4 (http://kpv.kazusa.or.jp/kpv4/) is an important tool to understand metabolic regulation and to generate hypotheses from publically available 'omics' data. The availability of genomics and transcriptomics information and combined analysis with metabolomics has facilitated the interpretation and understanding of complex biological processes [20].

Soybean is the most important legume crop, which provides sources of starch, dietary fiber, protein, lipids, and essential minerals oil and protein for human as well as for livestock. Soybeans are unique in legumes with contents of about 40% protein and 21% oil. It is the most widely grown protein/oilseed crop in the world, with both North and South America producing a large portion of the world's supply and constitutes about half of the global oilseed production [21]. Besides the consumption as a food and feedstock, soybean oil can potentially serve as a future fuel source with efforts being made to improve soy-diesel production [22], besides soybean protein-based biodegradable materials being researched can serve as an alternative for plastics [23]. Diverse soybean use makes it a widely desired crop plant, and the demand is rapidly increasing. Studies have demonstrated that genetics and environment can significantly affect soybean, with the production greatly influenced by climatic conditions, geographical location, and environmental stress such as drought, flooding, and salinity [24]. However, since complex interaction is involved

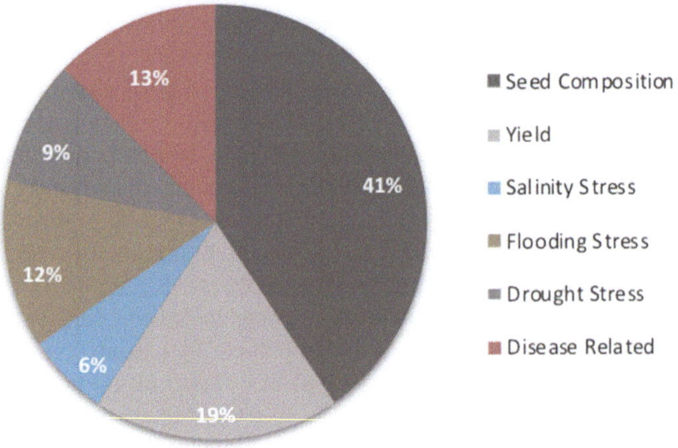

Fig. 2 Distribution of recent publications on applications of metabolomics in soybean by research area. Numbers of publications were obtained by searching for the corresponding keywords (e.g., metabolomics AND soybean seed) in PubMed with a time range of 2001–current (April 2018) and fractions were calculated by dividing the number obtained for each area by the total number of publications in metabolomics

between molecular, biochemical, and genetic mechanisms in soybean, a thorough understanding of the relationship between crop performance and metabolic response of soybean to various growing conditions is needed for its effective management. In recent years, metabolomics has been applied in soybean for various environmental stress tolerance enhancement (Fig. 2). Here in this chapter, we discuss recent progress in soybean metabolomics studies for biotic and abiotic stress tolerance and other growth conditions.

2 Metabolomics in Soybean Biotic/Abiotic Stress

Due to the high demand for soybean, quantity and quality control are critical for soybean production. In view of the demand, the environmental stress is one of the factors that can cause serious yield loss. Biotic or abiotic stresses could be defined as any alterations in the plant's growth conditions, which causes a change in its metabolic pathways to achieve a new state of homeostasis [25]. For example, several pathogens, such as bacteria, fungi, nematodes, virus, and insects or other environmental stresses, such as drought, heat, and salt are serious threats to soybean cultivation in the USA and other soybean growing countries, causing losses and increasing production costs every year. In order to improve soybean production and quality, the metabolomic analysis could help to monitor precursors, intermediates, and metabolic pathways to detect metabolic responses to various environmental stress. In soybean, several efforts have been made to improve salt tolerance [26], flooding [27], drought and heat stress [28], and various pathogen infections such as virus [29] and aphid [30] (Table 1) (Fig. 2).

A metabolomic study was performed to understand root nodulation by nitrogen-fixing symbiotic bacterium Bradyrhizobium japonicum in soybean. The study identified 2610 metabolites in root hairs using GC-MS and UPLC-QTOF-MS, and 166 metabolites were observed to be significantly associated with the responses to *B. japonicum* inoculation. Subsequent analysis revealed that trehalose was among the most strongly induced metabolites detected following *B. japonicum* inoculation. The study suggested that *B. japonicum* likely experiences osmotic stress during the infection process, either on the root hair surface or within the infection thread because of osmoprotectant trehalose [31].

Since wild soybean (*Glycine soja*) is considered to be resistant and adaptable to different environmental conditions, it has been utilized vastly in salt stress-related studies. Lu et al. (2013) characterized metabolic changes in cultivated (*Glycine max*) and wild soybean under salt stress. The results indicated that wild soybean contained higher sugar alcohols, disaccharides, and acetylated amino acids than the cultivated soybean, but with relatively lower amounts of monosaccharides, carboxylic acids, and unsaturated fatty acids. To obtain high yields in saline soil conditions, it is important to dissect the plant metabolic mechanisms in response to salt stress [26]b. Zhang et al. (2016) characterized 68 metabolites in cultivated soybean and wild soybean under neutral-salt and alkali-salt stresses using GC-MS to uncover the physiological and molecular differences in salt tolerance. The study revealed that wild soybean contained significantly higher amounts of phenylalanine, asparagine, citraconic acid, citramalic acid, citric acid, and α-ketoglutaric acid under neutral-salt stress, and higher amounts of palmitic acid, lignoceric acid, glucose, citric acid, and α-ketoglutaric acid under alkali-salt stress, than cultivated soybean resulting into lower inhibition in wild soybean than in cultivated soybean especially under alkali-salt stress [32]. Furthermore, another study corroborated the fact that wild soybean is more salt tolerant than semi-wild and cultivated soybeans because of higher amounts of specific metabolites in wild soybean [33]. Similarly, water stress was assessed in cultivated soybean (drought tolerant (NA5009RG) and sensitive (DM50048) genotypes) for overall growth, nitrogen fixation, ureide, and proline dynamics to elucidate metabolic changes in relation to the physiological responses in the nitrogen-fixing plants toward water limitation [34]. The results demonstrated differences in physiological responses between these two genotypes on the basis of metabolite variation in response to drought conditions. Moreover, drought/heat stress was further studied in cultivated soybean, and metabolomic data showed that the specific metabolites involved in various cellular processes, such as glycolysis, the tricarboxylic acid (TCA) cycle, the pentose phosphate pathway, and starch biosynthesis, were differentially accumulated in soybean leaves [28].

Also, metabolomics approach has been used to understand metabolism regulation in response to pathogen infections such as *Rhizoctonia solani* [35] and *Soybean Mosaic virus* [29] and supported the fact that change in particular carbohydrates and amino acids concentrations determine the plant response in stress conditions. With above-mentioned studies, thus, metabolomic profiling in soybeans demonstrated

Table 1 Details of significant efforts of metabolomics studies performed in soybean

Tissue	Purpose of the study	Methods	Metabolites	References
Leaf	Metabolome compartmentation assessment	NAQF and GC-MS	100	[41]
Seed	Comparative metabolite profiling for salt tolerance	HPLC-UV-ESI-MS	200	[26]a
Seed	Metabolite comparison in GM and conventional soybean	CE-ESI-TOF-MS	45	[42]
Germinated soybean	Defense response	LC-MS and GC-MS	–	[43]
Dried soybean	Fermentative capability of soybean starter	^1H NMR spectra	–	[44]
Root hairs	Metabolic response on *B. japonicum* infection	GC-MS, UPLC-QTOF-MS	199	[30]
Cotyledon	Defense-related prenylated isoflavones	LC-MS and NMR	26	[45]
Leaves and nodules	Water stress in tolerant and sensitive soybean	^1H NMR spectra	–	[34]
Leaf	Salt stress comparison in *G. max* and *G. soja*	GC–MS/LC–FT/MS	98	[26]b
Leaf	Aphid resistance in sensitive and tolerant soybean	CE–TOF-MS	196	[30]
Pods	Metabolite changes under greenhouse and field condition	GC-MS, LC-MS, and GC-FID	20	[36]
Leaf	Metabolic comparison in cultivated and semi-wild soybean	^1H NMR spectra	–	[46]

(continued)

Table 1 (continued)

Tissue	Purpose of the study	Methods	Metabolites	References
Seedling roots	Salt tolerance comparison in wild, semi-wild, and cultivated soybean	GC–MS/LC–FT/MS	47	[33]
Seedling	Shade tolerance at seedling stage	HPLC	12	[47]
Seed	Effect of shading on fatty acid metabolism	GC-MS	–	[48]
Leaf	Effect of drought and heat stress on sugar and nitrogen metabolism	GC-MS and LC-MS	266	[28]
Leaf and root	Metabolic response evaluation to Mg deficiency	GC-TOF	–	[39]
Seedling roots	Salt stress in *G. soja*	GC-MS	28	[26]c
Leaf	Effect of geographical dependence on metabolic composition	^{1}H NMR spectra	–	[38]

carbohydrates and amino acids concentrations associated with sugar and nitrogen metabolism are of prime significance under environmental stress conditions. Therefore, to combat environmental stress for increased productivity and yield, it is important to understand the key metabolite changes that regulate different physiological and biochemical processes in various stress conditions.

3 Metabolomics in Soybean Grown Under Different Growth Conditions

Since the metabolic compositions are likely affected by environmental conditions during growth, the effect of geographical location and various growth conditions on metabolites have been compared in soybean. Due to changing climate, reduced agriculture acreages, and increasing soybean demand, greenhouse cultivation can be helpful to enhance productivity. A greenhouse can provide extended growing seasons, minimal pest infestations, and protection against drastic weather fluctuations like frost, heat, rain, and wind. To understand metabolite changes in field and growth conditions, John et al. (2016) examined the metabolic responses of nine soybean varieties grown under field and greenhouse conditions. The metabolites such as glucose, sucrose, and pinitol and isoflavone aglycones were found to be increased, but the isoflavones glucoside content decreased in the greenhouse cultivated soybeans. The results indicated that genetic variation and growing conditions certainly influence the soybean metabolites composition [36]. Similarly, another study supported the metabolite composition change in various growth conditions [37]. Dwarf (Mini-Max) and commercially cultivated (Williams 82) soybean genotypes were assessed under greenhouse and field conditions. Interestingly, isoflavone aglycones content was observed higher in both genotypes in greenhouse grown samples as compared to the field grown samples. This study concluded that metabolite composition gets influenced by both growing conditions and cultivars [37]. Furthermore, the cultivated soybean (*G. max cv. Sinhwa*) was investigated by growing soybean in various geographical locations in Korea, namely the southernmost island and volcanic region of Korea, the central section and the limestone region of the Korean peninsula to observe the metabolite composition of soybean leaves under various environmental conditions. Due to differences in soil compositions and climatic conditions between the two growing areas, differences in accumulations of pinitol and diverse flavonoids were noted in soybean leaves, suggesting that soybean plants have the distinct metabolism and physiological adaptation toward different environmental conditions [38].

Besides metabolomics being used for understanding environmental factors and their impact on soybean, it is also being used to study the impact of mineral deficiency. For example, magnesium (Mg) deficiency has been reported to affect soybean plant growth and development through regulating key metabolic pathways [39]. The results showed that Mg deficiency could cause large metabolite alterations in carbon and nitrogen metabolism. The Mg-deficient plants displayed distinct changes at a metabolic level from controls at 4 days after stress (DAS) and were clearly discriminated at 8 DAS. Also, the carbon and nitrogen metabolic responses were found to be distinct in leaves and roots under Mg deficiency. The data suggested that most amino acids (such as phenylalanine, asparagine, leucine, isoleucine, glycine, glutamine, and serine) had higher accumulation in the leaves, while most organic acids (including pyruvic acid, citric acid, 2-ketoglutaric acid, succinic acid, fumaric acid, and malic acid) were significantly decreased in the roots [39]. From the studies mentioned above, it can be understood that the complexity of plant physiological processes necessitates the use of metabolomics analysis for a comprehensive understanding with regard to metabolite regulation and response to various environmental and genetic factors.

4 Integration of Metabolomics with Other Omics Approaches in Soybean

Advancements in various omics approaches and computational tools have provided information related to gene function, genome structures, biological pathways, metabolic and regulatory networks. Although there are many biological network models available, they do not always provide a complete depiction of cellular and molecular networks of the plant systems solely based on genome, transcriptome, or metabolome data. Thus, integration of multiple 'omics' approaches is essential to decipher the physiological, biochemical, and molecular levels to develop high-quality soybean with stable yield.

In soybean, a combination of approaches has led to successful discoveries, for instance, Kovinich et al. (2011) combined gene expression and metabolite data to elucidate the control of the R locus identification of pigment biosynthesis genes [40]. Recently, proteomics and metabolomics were integrated to investigate the changes in soybean leaf metabolism upon ethylene (ET), abscisic acid (ABA), and combined ABA + ET treatments [19]. A total of 4129 unique protein groups were identified. Further analysis showed an abundance of specific proteins associated with flavonoid and isoflavonoid biosynthesis in response to ET treatment. In addition, metabolome analysis assigned 79 metabolites, which further corroborated the accumulation of flavonoids and isoflavonoids upon ET treatment [19]. Hence, above-mentioned studies clearly illustrated that a combination of approaches is required for better understanding of physiological and biochemical processes in soybean.

5 Conclusion and the Future Outlook

The advancements in metabolomics technologies have accelerated the possibility of biological process regulations study at different levels for the desired traits in soybean. Recent developments in computational resources, statistical tools, and high-throughput instruments have paved the way for the discovery of novel metabolites and specific metabolite networks toward improving soybean seed composition and environmental stress tolerance, etc. These advances would greatly help in the precise understanding of various biological processes and signaling pathways. Nevertheless, challenges remain in multiple areas.

Although there are several metabolomics platforms and resources available, the greatest challenge in metabolomics is still to decide the best approach in experiments to achieve best biological insight. In addition, there are thousands of plant metabolites, out of which many are still unknown. Therefore, no single method is sufficient to evaluate plant metabolomics, a combination of approaches (i.e., LC-MS, GC-MS, NMR) dependent on the chemical properties of the metabolite could be implemented to acquire comprehensive metabolome information.

Due to the complex interaction between several traits, it is also essential to combine metabolomics data with other 'omics' databases (i.e., genomics, transcriptomics, and proteomics) to obtain a clear picture of the biological processes during soybean growth, development, and stress conditions. Despite the availability of metabolomics resources and other omics technologies, computational tools and database still need to be designed for the utilization of publicly available data. Therefore, the establishment of high-throughput data analysis and processing standards will be crucial step to further advance the multi-omics approach for more comprehensive understanding.

In spite of these challenges, efforts are underway to create better approaches and protocols to make advances in the metabolomics field. Even though the future outlook for metabolomics is promising, the currently available technologies are fairly progressed allowing identification of unknown genes, biomarker discovery, and rapid discovery of novel metabolites. The information presented in this chapter will provide glimpses of the current scenarios and future perspectives for the stress tolerance improvement of soybean.

Acknowledgements The authors would like to thank Dr. Gunvant Patil and Dr. Humira Sonah for critical review and manuscript editing.

Conflicts of interest The authors declare no conflict of interest.

References

1. Fiehn, O.: Plant Mol. Biol. **48**, 155–171 (2002)
2. Hall, R.D., Brouwer, I.D., Fitzgerald, M.A.: Physiol. Plant. **132**, 162–175 (2008)
3. Hong, J., Yang, L.T., Zhang, D.B., Shi, J.X.: Int. J. Mol. Sci. **17** (2016)
4. Trethewey, R.N., Krotzky, A.J., Willmitzer, L.: Curr Opin Plant Biol **2**, 83–85 (1999)
5. Griffin, J.L.: Philos. Trans. R. Soc. B-Biol. Sci. **361**, 147–161 (2006)
6. Fiehn, O., Kopka, J., Dormann, P., Altmann, T., Trethewey, R.N., Willmitzer, L.: Nat. Biotechnol. **18**, 1157–1161 (2000)
7. Madalinski, G., Godat, E., Alves, S., Lesage, D., Genin, E., Levi, P., Labarre, J., Tabet, J.C., Ezan, E., Junot, C.: Anal. Chem. **80**, 3291–3303 (2008)
8. Pluskal, T., Castillo, S., Villar-Briones, A., Oresic, M.: BMC Bioinformatics **11**, 395 (2010)
9. Tsugawa, H., Cajka, T., Kind, T., Ma, Y., Higgins, B., Ikeda, K., Kanazawa, M., VanderGheynst, J., Fiehn, O., Arita, M.: Nat. Methods **12**, 523–526 (2015)
10. (a) Smith, C.A., Want, E.J., O'Maille, G., Abagyan, R., Siuzdak, G.: Anal. Chem. **78**, 779–787 (2006); (b) Mahieu, N.G., Genenbacher, J.L., Patti, G.J.: Curr. Opin. Chem. Biol. **30**, 87–93 (2016)
11. a) Sturm, M., Bertsch, A., Gropl, C., Hildebrandt, A., Hussong, R., Lange, E., Pfeifer, N., Schulz-Trieglaff, O., Zerck, A., Reinert, K., Kohlbacher, O.: BMC Bioinformatics **9**, 163 (2008); b) Rost, H.L., Sachsenberg, T., Aiche, S., Bielow, C., Weisser, H., Aicheler, F., Andreotti, S., Ehrlich, H.C., Gutenbrunner, P., Kenar, E., Liang, X., Nahnsen, S., Nilse, L., Pfeuffer, J., Rosenberger, G., Rurik, M., Schmitt, U., Veit, J., Walzer, M., Wojnar, D., Wolski, W.E., Schilling, O., Choudhary, J.S., Malmstrom, L., Aebersold, R., Reinert, K., Kohlbacher, O.: Nat. Methods **13**, 741–748 (2016)
12. a) Spicer, R., Salek, R.M., Moreno, P., Canueto, D., Steinbeck, C.: Metabolomics **13**, 106 (2017); b) Tsugawa, H.: Curr. Opin. Biotechnol. **54**, 10–17 (2018)
13. Sucaet, Y., Wang, Y., Li, J., Wurtele, E.S.: BMC Bioinformatics **13**, 267 (2012)

14. Qiu, F., Fine, D.D., Wherritt, D.J., Lei, Z., Sumner, L.W.: Anal. Chem. **88**, 11373–11383 (2016)
15. a) Malongane, F., Mcgaw, L.J., Nyoni, H., Mudau, F.N.: Food Chem. 257, 90–100 (2018); b) Ossa, D.E.H., Gil-Solsona, R., Penuela, G.A., Juan, V., Hernandez, F.J.: Food Chem. **250**, 89–97 (2018)
16. Yonekura-Sakakibara, K., Saito, K.: Biotech. Lett. **28**, 1983–1991 (2006)
17. a) Horikoshi, H.M., Sekozawa, Y., Kobayashi, M., Saito, K., Kusano, M., Sugaya, S.: Plant Physiol. Biochem. **126**, 134–141 (2018); b) Rajniak, J., Giehl, R.F.H., Chang, E., Murgia, I., von Wiren, N., Sattely, E.S.: Nature Chem. Biology **14**, 442–+ (2018)
18. Zivy, M., Wienkoop, S., Renaut, J., Pinheiro, C., Goulas, E., Carpentier, S.: Frontiers Plant Sci. **6** (2015)
19. Gupta, R., Min, C.W., Kramer, K., Agrawal, G.K., Rakwal, R., Park, K.H., Wang, Y., Finkemeier, I., Kim, S.T.: Proteomics **18** (2018)
20. a) Fernie, A.R., Schauer, N.: Trends in Genetics, 25, 39–48 (2009); b) Saito, K., Matsuda, F.: Annu. Rev. Plant Biol. **61**, 463–489 (2010)
21. Chaudhary, J., Patil, G.B., Sonah, H., Deshmukh, R.K., Vuong, T.D., Valliyodan, B., Nguyen, H.T.: Frontiers Plant Sci. **6** (2015)
22. Candeia, R.A., Silva, M.C.D., Carvalho, J.R., Brasilino, M.G.A., Bicudo, T.C., Santos, I.M.G., Souza, A.G.: Fuel **88**, 738–743 (2009)
23. Song, F., Tang, D.L., Wang, X.L., Wang, Y.Z.: Biomacromol **12**, 3369–3380 (2011)
24. Deshmukh, R., Sonah, H., Patil, G., Chen, W., Prince, S., Mutava, R., Vuong, T., Valliyodan, B., Nguyen, H.T.: Frontiers Plant Sci. **5**, (2014)
25. Shulaev, V., Cortes, D., Miller, G., Mittler, R.: Physiol. Plant. **132**, 199–208 (2008)
26. a) Wu, W., Zhang, Q., Zhu, Y.M., Lam, H.M., Cai, Z.W., Guo, D.J.: J. Agric. Food Chem. **56**, 11132–11138 (2008); b) Lu, Y.H., Lam, H.M., Pi, E.X., Zhan, Q.L., Tsai, S.N., Wang, C.M., Kwan, Y.W., Ngai, S.M.: J. Agric. Food Chem. **61**, 8711–8721 (2013); c) Jiao, Y., Bai, Z.Z., Xu, J.Y., Zhao, M.L., Khan, Y., Hu, Y.J., Shi, L.X.: Plant Physiol. Biochem. 126, 187–196 (2018)
27. Wang, X., Zhu, W., Hashiguchi, A., Nishimura, M., Tian, J.K., Komatsu, S.: Plant Mol. Biol. **94**, 669–685 (2017)
28. Das, A., Rushton, P.J., Rohila, J.S.: Plants (Basel) **6**, (2017)
29. Diaz-Cruz, G.A., Cassone, B.J.: Physiol. Mol. Plant Pathol. **102**, 79–87 (2018)
30. Sato, D., Akashi, H., Sugimoto, M., Tomita, M., Soga, T.: J. Chromatogr. B-Anal. Technol. Biomed. Life Sci. **925**, 95–103 (2013)
31. Brechenmacher, L., Lei, Z.T., Libault, M., Findley, S., Sugawara, M., Sadowsky, M.J., Sumner, L.W., Stacey, G.: Plant Physiol. **153**, 1808–1822 (2010)
32. Zhang, J., Yang, D., Li, M., Shi, L.: PLoS ONE **11**, e0159622 (2016)
33. Li, M., Guo, R., Jiao, Y., Jin, X., Zhang, H., Shi, L.: Front Plant Sci. **8**, 1101 (2017)
34. Silvente, S., Sobolev, A.P., Lara, M.: PLoS ONE **7**, e38554 (2012)
35. Copley, T.R., Duggavathi, R., Jabaji, S.: Plos One, **12**, (2017)
36. John, K.M.M., Natarajan, S., Luthria, D.L.: Food Chem. **211**, 347–355 (2016)
37. John, K.M.M., Khan, F., Luthria, D.L., Matthews, B., Garrett, W.M., Natarajan, S.: J. Food Biochem. **41**, (2017)
38. Yun, D.Y., Kang, Y.G., Kim, E.H., Kim, M., Park, N.H., Choi, H.T., Go, G.H., Lee, J.H., Park, J.S., Hong, Y.S.: Food Res. Int. **106**, 842–852 (2018)
39. Yang, N., Jiang, J.L., Xie, H.L., Bai, M.Y., Xu, Q.Z., Wang, X.G., Yu, X.M., Chen, Z.C., Guan, Y.F.: Frontiers in Plant Sci. **8**, (2017)
40. Kovinich, N., Saleem, A., Arnason, J.T., Miki, B.: Bmc Genomics **12**, (2011)
41. Benkeblia, N., Shinano, T., Osaki, M.: Metabolomics **3**, 297–305 (2007)
42. Garcia-Villalba, R., Leon, C., Dinelli, G., Segura-Carretero, A., Fernandez-Gutierrez, A., Garcia-Canas, V., Cifuentes, A.: J. Chromatogr. A **1195**, 164–173 (2008)
43. Choi, J.N., Kim, J., Lee, M.Y., Park, D.K., Hong, Y.S., Lee, C.H.: J. Agric. Food Chem. **58**, 4258–4267 (2010)
44. Ko, B.K., Kim, K.M., Hong, Y.S., Lee, C.H.: J. Agric. Food Chem. **58**, 8738–8747 (2010)
45. Cheng, J.Y., Yuan, C.H., Graham, T.L.: Phytochemistry **72**, 875–881 (2011)

46. Yun, D.Y., Kang, Y.G., Yun, B., Kim, E.H., Kim, M., Park, J.S., Lee, J.H., Hong, Y.S.: J. Agric. Food Chem. **64**, 5773–5783 (2016)
47. Liu, J., Hu, B.Y., Liu, W.G., Qin, W.T., Wu, H.J., Zhang, J., Yang, C.Q., Deng, J.C., Shu, K., Du, J.B., Yang, F., Yong, T.W., Wang, X.C., Yang, W.Y.: Sci. Rep. **7**, (2017)
48. Yang, C.Q., Iqbal, N., Hu, B.Y., Zhang, Q.H., Wu, H.J., Liu, X., Zhang, J., Liu, W.G., Yang, W.Y., Liu, J.: Anal. Methods **9**, 2144–2152 (2017)

Phyto-Stem Cell: New Door in Cosmetology

Megha Jha

Abstract The recent upsurge in use of herbal medicines and cosmetics is based on new technical inputs that include the use of biologically active ingredients, genetic study for individual skin care, regenerative therapies, and tissue engineering. The biological active ingredients not only use in skin care products but also for use in various dermatological disorders. A healthy balanced mind and body are associated with muscle, bone, skin, hair, nails, subcutaneous tissues, and glands. The stem cells can be nudged to develop into specialized cell types to regenerate tissue. Cosmetics nowadays take advantage from those technologies for achieving healthy results. One of them is stem cell technology; the idea behind using stem cells is that they stimulate the growth of more stem cells in skin. Plant stem cell extracts act as an active ingredient in cosmetics due to the biologically beneficial properties; therefore, some researchers have focus toward stem cells for cosmetic formulation preparations. There is a concern about harmful chemical used in personal care products which has increased consumer interest in naturally prepared cosmetics formulations. Many biotech companies are manufacturing anti-aging, skin soothing, acne-prone personal care products from biologically active ingredients derived from plants. The present review focuses on biotechnologically prepared plant stem cells for production of active ingredients in cosmetic industry for keeping skin looking youthful.

Keywords Subcutaneous · Acne prone · Cosmetology

1 Introduction

The term cosmeceuticals was first used by Raymond Reed founding member of U.S. Society of Cosmetics Chemist in 1961. He actually used the word to brief the active and science based cosmetics. The above term cosmetic and therapeutic was further used by Dr. Albert Kligman in 1984 [1]. In the last few decades, there has been an

M. Jha (✉)
Pinnacle Biomedical Research Institute, Bhopal, Madhya Pradesh, India
e-mail: meghajhabtbpl@gmail.com

M. Khoobchandani and A. Saxena (eds.), *Biotechnology Products in Everyday Life*, EcoProduction, https://doi.org/10.1007/978-3-319-92399-4_13

exponential growth in the field of herbal cosmeceuticals which are gaining popularity in world market because of their skin friendliness, purity, and less side effects. In India, cosmetic industry has a plethora of natural cosmetic brands as Himalaya, Biotique, Patanjali, Dabur, VLCC, Lotus, etc.

Herbal cream formulation is very popular and is said to have various beneficial effects such as soothing of skin irritation as it contains powerful antioxidants that help improve skin texture, along with the benefits of Vitamins C and E and fatty acids, which promoting relaxation, offering intense hydration, and moreover it also helps in reducing the appearance of the first signs of aging as elastic and toning of skin. The phytocomponents play a major role in healing potential as described in Ayurvedic literature resources and characterized in Ayurvedic pharmacopeias. During manufacturing, cosmetic formulations were standardized at all quality testing levels of manufacturing from plant selection, procurement, molecular biomarkers mechanistic level to product development.

Herbal cosmetics and ingredients do not require having Food and Drug administration approval before they reach market. Like cosmetics, these are only subjected for their safety assessment as per the given rules of FDA. Generally, it is not mandatory for a manufacturer to claim about product formulation [2]. There are rising numbers of consumers concerned about the ingredients such as synthetic chemicals, mineral oils, who are demanding more natural products with traceable and more natural ingredients free from harmful chemicals [3].

Past literature confirmations about the use of phyto-stem cells technology for slow skin aging and keep skin youthful for long time. Phyto-stem-cell-based cosmetics are particularly preferred because of their eco-friendly nature, and it would not provide any side effects to the skin and the body. Several plants extracts are used for new cosmetic products development. Phyto-stem cells products are the formulations, which represent cosmetic associated with active phytoconstituents; thus, it helps to reduce damage, irritation, and aging of the skin (Fig. 1).

2 Plant Cell Culture Technology

Plant tissue culture allows the propagation of undifferentiated plant cells to regenerate a whole plant or to produce single cells culture for production of secondary metabolites [5]. The use of plant cell culture allows a controlled production independent of the season and weather influences, also free from the risk of contamination from pesticides. Cosmetologists are desperately looking for new innovative natural approaches for cosmetic applications by the help of plant tissue culture technique. Cosmetic societies are bringing up new aspect of cosmetic plant-derived metabolites.

Plant cell culture technology should ensure the growth of plant cell in aseptic condition. The process involves the selection of appropriate plant material and sterilization of a tissue to eliminate damage of meristem cell which is required for new cell line creation. Sterilized plant tissue is reduced to fragments (explants) and placed on the petri plates containing solid medium. An unorganized mass produced (callus)

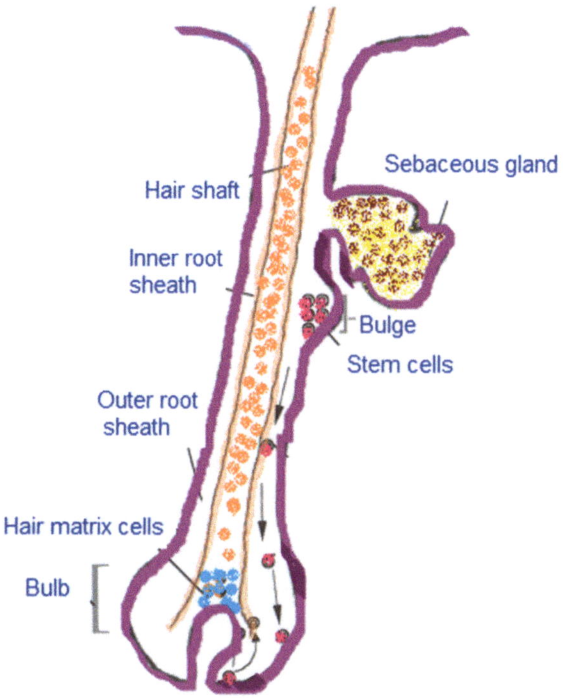

Fig. 1 Schematic structure of hair follicle. Modified from [4]

which must be regularly transferred to a proper nutrient medium including a source of organic carbon and energy (most frequently a saccharose) and plant hormones equal proportions (auxin and cytokine), vitamins, and also micro- and macroelements. Plants can be induced by hormones auxin to cytokinin ratio greater than cause root development on many replicate plantlets, and treatment with auxin to cytokinin ratio is lesser than induces shoot development on many replicate plantlets. The suspension cultures require gradual adaptation for growth inside a conical flask (200 mL) and then to grow inside bioreactors (100 L). Cultures carried in bioreactors must have ensured a constant temperature and proper gaseous exchange level required for cell metabolism. The increase in biomass is monitored by measurement of: sugar contents, electrical conductivity, pH level, optical density measurement, cell vitality, and contents of secondary metabolites. The choice of the suitable cell lines depends on the large-scale production of cultures and lesser doubling time. The established suspension cell culture processed with high-pressure treatment to break the suspended cells in a mixture thus releases the active ingredients. The extract obtained from plant stem cells can be then further encapsulated for better flux and increased permeability in various nano-structured carrier systems for topical delivery as a cosmetic product [6].

3 Phyto-Stem Cells

Stem cells are characterized by the ability to regenerate and continuously give rise to new specialized cells that have the potential to grow into any organ, tissue, or cell in the plant's body. Phyto-stem cells do not undergo the aging process, and their everlasting life gives rise to specialized and unspecialized cells. They have the potential to grow into any organ, tissue, or cell in the body and, by being multi-functional, they create a powerful nutrient balm; a wealth of amino acids, sugars, and proteins that you are extracting and putting into the product. The use of phyto-stem cell is to create secondary metabolite by tissue culture technique as it has several advantages, such as availability of fresh materials, standardization, high levels of consistency from batch to batch, extracted compound are safe as produced in aseptic conditions, cells can be induced to produce specific compounds of interest. That is the brief by which scientists have more control over the quality and quantity of anti-aging substance the plant produces.

The stem cells from which the plant body develops are located in shoot and root meristems, and new research indicates that the balance between proliferation and differentiation in each is controlled by related proteins that interact in a similar feedback network. By using cell cultures, plant biotech industries can grow wide variety of plants, creating a sustainable registry of plant extracts without disrupting the environment (Fig. 2).

Extensive effort has been expanded by pharmaceuticals and cosmetic industries to understand the skin aging processes. Stem cells have attracted similar interest from the cosmetics industry, and various companies are already exploring their potential for skin care products. In the entire life cycle of plants, stem cells remain active over hundreds to thousands of years, revealing the exquisite precision in the underlying control of proliferation, self-renewal, and differentiation (Fig. 3).

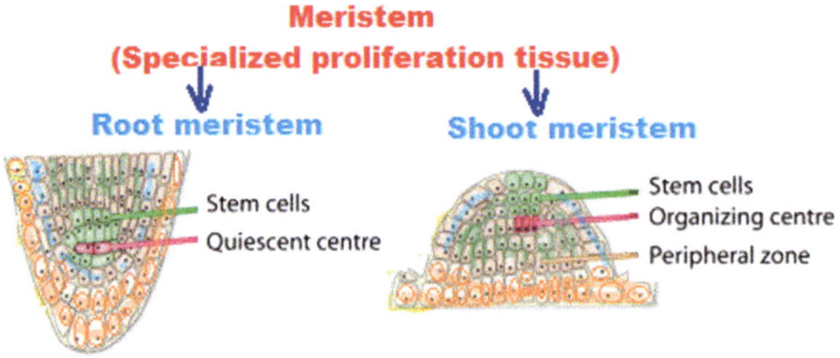

Fig. 2 Diagrammatic representation of stem cells in plant meristem (root and shoot meristem)

Fig. 3 Flow diagram of Phyto-stem cell technology

4 Genetic of Healthy Skin

Aging depends on various factors such as diet, environment, genetic factors, and other personal habit. It is a multifaceted progression of the skin layers: epidermis, dermis, and subcutis with denaturing proteins and reduced functioning of regenerative cells. Stem cell residing in the epidermis is divided into basal layer stem cells and inter-follicular stem cells [7]. A decline in epidermal stem cells has been observed with reference to shortened telomeres, which reduced proliferative potential when cells exposed to UV radiations thus, leading to aging and also cell death [8].

A rejuvenating enzyme telomerase can add telomeric repeats at the end of the chromosome and these repeats with the help of telomerase enzyme prevent the deterioration as well as shortening of chromosomal end each time after the cell division. Since, telomere is an important tool for maintaining chromosomal stability. Telomerase consist of Protein and RNA sequences as a template for synthesizing telomere DNA.

5 Benefits of Plant Stem Cells for Human Skin

Natural cosmetic is general term applied to all preparation and external conditioning or beautifying the body [9]. Stem cells play an important role in skin care as they are having self-renewal property. Antioxidants are considered important cosmeceuticals on justification of skin care benefits by protecting cell from UV damage and by neutralizing free radicals. A world-class cosmetic product contains plant extracts derived from plant stem cells and that helps in protecting the skin from damage. Biotechnology is no longer limited to cloning techniques. Through stem cell technology, researchers are able to extract tissues from plants. The plant kingdom has provided an endless source of medicinal plants first used in their crude forms as herbal ointments and powders. Today, researcher focuses upon the plants ability to regenerate stem cell for human use. Hence, it is believed that a nature has provided an excellent storehouse of remedies to cure skin diseases. Stem cell technology holds promise for stem cell cosmetics, which involves plant extracts for skin fairness as well as reducing side effects. Research shows that plant stem cells can slow skin aging by defending against extrinsic stress, keeping skin looking youthful, longer. This discovery opened the door in cosmetic stem cell research.

6 Phyto-Stem Cells

6.1 Edelweiss Stem Cells

Edelweiss (*Leontopodium alpinum*) is tiny flower that withstand harsh alpine climes, and its extract is having a strong antioxidative and radical scavenging activity as it contains leontopodic acids A and B, an strong antioxidants. An Italian company, the Institute of Biotechnological Research, that produces the edelweiss stem cells explains an interesting technique in which they chop up some of the plant's biomass, and the surrounding cells revert to stem cells in order to protect the plant with a wound healing tissue called callus. Edelweiss also contains various effective components such as polyphenols, belongs to classes of phenylpropanoids (phenolic acids, glycosides, flavonoids, coumarines, and lignans), terpenes (sesquiterpenes and diterpene acids), and alkaloids (benzofuran and pyrane derivatives) [10]. According to pharmacological aspects, the extract of Edelweiss from aerial parts and roots possessing an anti-inflammatory and antibacterial property due to the presence of phenylpropanoids [11]. The analytical chemistry of Edelweiss revealed the presence of glycosides and aglycones of flavonoids (luteolin, quercetin, and apigenin) as well as high content of leontopodic, chlorogenic, and 3, 5-dicaffeoylquinic acids, that may help to lead younger-looking skin. It is said to have cytoprotective property and allegedly prevents collagen breakdown [12].

6.2 Apple Stem Cells

PhytoCellTec *Malus Domestica* is a liposomal preparation based on the rare Swiss apple stem cell called *Uttwiler Spatlauber*. Plant stem cells can be integrated easily into cosmetic products in form of artificial vesicles to enhance anti-aging property of skin. It not only protects skin's own stem cells but has been shown to have excellent age-delaying and anti-wrinkle properties, and at this time it is one of the most pioneering ingredients in skin care [13].

6.3 Lilac Stem Cells

Syringa vulgaris is a shrub commonly known as Lilac. Studies are showing that the enzyme produced by plant tissue culture reduces acne lesions up to 40%. Tissue culture technique also helps to slow down the sebum production, and it is a tyrosinase inhibitor. Enzyme tyrosinase—a key factor in the melanogenesis process—ultimately leads to the deposition of hyper-pigmentation. Past research revealed that the uses of lilac stem cells in skin lightening products have beneficial effect in post inflammatory hyper-pigmentation, and its results recommend that it has an antioxidant, anti-inflammatory properties, and reduces the trans-epidermal water loss. Verbascoside is the active component, an effective antioxidant, anti-inflammatory agent, and anti-acne also. Verbascoside has also been shown to suppress the 5-α reductase enzyme that triggers sebum activity—one of the four contributing factors in the pathogenesis of acne. Lilac stem cells inhibit 5α-reductase, an enzyme involved in sebum production, and they also decrease the pro-inflammatory chemokine IL-8. In a clinical research study, lilac stem cell extract reduced lesions in 29 acne patients by 40% in 30 days and significantly decrease in inflammation and melanin pigmentation [14].

6.4 Argan Stem Cell

The argan tree recognized as *Argania spinosa* is endemic to semi-desert areas of southwestern Morocco, where they cover more than 8000 km^2 and have important socioeconomic and environmental impacts. Active ingredients of *Argania spinosa* were tocopherol and essential fatty acids (linoleic acid and oleic acid). Argan is capable of both protecting and vitalizing human skin stem cells. Studies revealed that it accelerates the natural repair process and helps to tighten, tone, revitalizing and protecting the skin cells. Previous studies from the manufacturers revealed that argan is working in the skin cells. It reduces the wrinkle depth by up to 26% and increases skin density, encouraging collagen and elastin growth [15].

6.5 *Echinacea* Stem Cell

Echinacea Stems GX™ is plant-based stem cell obtained from the culture of *Echinacea angustifolia,* and it is traditionally used in neutraceuticals as an immunostimulant ingredient. Phenylpropanoid such as echinacoside, an active ingredient in Echinacea Stems GX™, efficiently preserves the capillary wall and reinforces the micro-vascular environment. The skin around the eyes is the thinnest body skin and is exposed to a wide range of stress conditions, such as aging. These external factors cause an inflammatory response to activate macrophages that release a large amount of nitric oxide, a potent vasodilator. Thus, attack the capillary wall and improve capillary permeability that causes blood vessel leakage, and the hemoglobin molecule is degraded into bilirubin and biliverdin responsible for skin darkens. Echinacea Stems GX™ strengthens the thin and delicate eye contour skin and stimulate the immune system, suppresses infection also have regenerating, wound healing, anti stress, anti-inflammatory, antioxidant, skin revitalizer property [16].

6.6 Grape Stem Cell

Grape (*Vitis vinifera*), one of the most commonly consumed fruits in the world, contains a variety of active compounds, including organic acids, oils, and polyphenols. Melatonin (N-acetyl-5-methoxytryptamine) a molecule with low molecular weight and an indole-based structure is an important compound present in grapes. Melatonin is a potent free radical scavenger and suppresses UV-induced damage to skin cells [17]. Phenolics are derivatives of hydroxycinnamic acid and hydroxybenzoic acid and polyphenols include flavonoids, stilbenes, and proanthocyanidins [18, 19]. The grape stem cells contain special epigenetic factors and metabolites which are able to protect human skin stem cells against UV radiation [20].

6.7 Mushroom Stem Cells

Mushrooms are spore-bearing fruiting body belonging to phylum Ascomycota and Basidiomycota. Every cell formed by a fungus can function as a "stem cell." Each adult stem cell from a mushroom is totipotent [21, 22]. Mushrooms are rich in protein, vitamins, minerals,, and excellent sources of β-glucan, selenium, thiamine, riboflavin, niacin, pantothenic acid, and folic acid, etc. [23, 24]. Since mushrooms contain valuable ingredients, they are helpful in maintaining one's optimal weight, longevity, and unnecessary aging [25, 26].

Nowadays, formulation was prepared using mushroom for topical application in the form of gels, creams, ointments, and other facial cosmetic products. These mushrooms include Shiitake (*Lentinula edodes*), Maitake (*Grifola frondosa*), Reishi or

Lingzhi (*Ganoderma lucidum*), Fu Ling (*Wolfiporia extensa*), Yartsa Gunbu (*Cordyceps sinensis*), cauliflower mushroom (*Sparassis latifolia*, formerly Sparassis crispa), and jelly fungi (Tremella spp.) [27–29]. Mushroom stem cells are being extensively studied by many researchers for skin cancer, healing, and angiogenesis.

7 Advantages of Plants Based Cosmetics Over Synthetic

Currently, much of the scientific research focuses on the use of natural cosmetics which was prepared by top 10 leading brands such as LOTUS, Himalaya, Biotique, Shahnaz Husain, Hindustan Unilever, VLCC, Emami. However, many researchers acknowledge the potential benefits of their use in personal care products. The aim of developing effective product is recognizing that the various plants have different properties and metabolism. Nowadays, plant stem cell-based OTC products are gaining popularity and public prefer natural products over chemicals for their personal care to enhance their beauty as these products supply the body with nutrients and enhance health by providing satisfaction as these are free from the synthetic harmful chemicals used in cosmetics.

8 Conclusion

Nowadays, society has an increasing interest in naturally prepared cosmetics formulations. Several researches support the beneficial effect of extracts, such as antioxidant capacity, tyrosinase inhibition, and antimicrobial activity. The present review refers to use of stem cell technology for preparation of healthy topical formulations. So, scientific studies aiming at the development, evaluation, and application of plant stem cell in topical formulations and that simultaneously meet consumer concerns challenges. Natural products are mild and biodegradable, exhibiting low toxicity and minus side effects. Therefore, phyto-stem cell produced by biotechnology and encapsulated for better properties as they were compatible to human skin and support skin stem cells protecting skin from UV damage. Stem cells contain growth-regulating factors, which play a role in cell division, growth of new cells, and for the production of collagen and proteins. If compared to synthetic low-grade cosmetic products, plant origin cosmetics are safe to use. As phyto-stem cells are made of natural ingredients, people do not have to worry about getting skin rashes or itchiness.

References

1. Saha, R.: Cosmeceuticals and herbal drugs: practical uses. Int. J. Pharm. Res. Sci. **3**, 59–65 (2012)
2. Kapoor, V.P.: Herbal cosmetics for skin and hair care. Nat. Prod. Radiance **4**(4), 306–314 (2005)
3. Escamilla, M., Ferre, A., Hidalgo, C., Fuentes, N., Kaps, R.: Revision of european ecolabel criteria for soaps, shampoos and hair conditioners. Joint Res. Centre Eur. Comm. pp. 1–40 (2012)
4. Niezgoda, A., Niezgoda P., Nowowiejska L., Białecka A., Męcińska-Jundziłł K., Adamska U., and Czajkowski R.: Properties of skin stem cells and their potential clinical applications in modern dermatology. Eur. J. Dermatol. **27**(3), 227–236 (2017)
5. Moscatiello, R., Baldan, B., Navazio, L.: Plant cell suspension cultures. Methods Mol. Biol. **953**, 77–93 (2013)
6. Morus, M., Baran, M., Rost-Roszkowska, M., Skotnicka-Graca, U.: Plant stem cells as innovation in cosmetics. Acta Pol. Pharm. **71**(5), 701–707 (2014)
7. Blanpain, C., Fuchs, E.: Epidermal stem cells of the skin. Annu. Rev. Cell Dev. Biol. **22**, 339–373 (2006)
8. Shen, Q., Jin, H., Wang, X.: Epidermal stem cells and their epigenetic regulation. Int. J. Mol. Sci. **14**(9), 17861–17880 (2013)
9. Mushtaq, A., Khan, M.A., Zafar, M.: Traditional herbal cosmetics used by local women communities in district Attock of Northern Pakistan. Indian J. Tradit. Knowl. **7**(3), 421–424 (2008)
10. Schwaiger, S., Cervellati, R., Seger, C., Ellmerer, E.P., Nancy, A., Isabelle, R., Godenir, C., Andre, P., Gafner, F., Hermann, S.: Leontopodic acid—a novel highly substituted glucaric acid derivative from edelweiss (*Leontopodium alpinum* Cass.) and its antioxidative and DNA protecting properties. Tetrahedron **61**(19), 4621–4630 (2005)
11. Dobner, M.J., Sosa, S., Schwaiger, S.: Anti-inflammatory activity of Leontopodium alpinum and its constituents. Planta Med. **70**(6), 502–508 (2004)
12. Schwaiger, S., Seger, C., Wiesbauer, B., Schneider, P., Ellmerer, E.P., Strum, S.: Development of an HPLC-PAD-MS assay for the identification and quantification of major phenolic edelweiss (*Leontopodium alpium* Cass.) constituents. Phytochem. Anal. **17**(5), 291–298 (2006)
13. Schmid, D., Schurch, C., Blum, P., Belser, E., Zulli, F.: Plant stem cell extract for longevity of skin and hair. Int. J. Appl. Sci. **134**(5), 30–35 (2008)
14. Price, C.: An in-depth look at lilac stem cell extract in skin care. Pract. Dermatol. pp. 46–48 (2014)
15. Justamante, M.S., Báñez, S., Villanova, J., Pérez-Pérez, J.M.: Vegetative propagation of argan tree (*Argania spinosa*(L.) Skeels) using in vitro germinated seeds and stem cuttings. Sci. Hortic. **225**(18), 81–87 (2017)
16. Morus, M., Baran, M., Rost-Roszkowska, M., Skotnicka-Graca, U.: Plant stem cells as innovation in cosmetics. Acta Pol. Pharm. Drug Res. **71**(5), 701–707 (2014)
17. Fiume, M.M., Bergfeld, W.F., Belsito, D.V., Hill, R.A., Klaassen, C.D., Liebler, D.C., Marks, J.G., Shank, R.C., Slaga, T.J., Snyder, P.W., Andersen, F.A.: Safety assessment of *Vitis vinifera* (Grape)-derived ingredients as used in cosmetics. Int. J. Toxicol. **33**(6), 48–83 (2014)
18. Delmas, D.: Resveratrol: Sources, Production and Health Benefits. Nova Science Publishers Inc., Hauppauge (2013)
19. Pezzuto, J.M., Kondratyuk, T.P., Ogas, T.: Resveratrol derivatives: a patent review (2009–2012). Expert Opin. Ther. Pat. **23**, 1529–1546 (2013)
20. Soto, M.L., Falqué, E., Domínguez, H.: Relevance of natural phenolics from grape and derivative products in the formulation of cosmetics. Cosmetics **2**, 259–276 (2015)
21. Cheung, P.C.: Mushrooms as Functional Foods, pp. 1–34. Wiley, Hoboken (2008)
22. Miles, P.G., Chang, S.T.: Mushrooms: Cultivation, Nutritional Value, Medicinal Effect, and Environmental Impact, 2nd edn, pp. 1–26. CRC Press, Boca Raton (2004)
23. Poucheret, P., Fons, F., Rapior, S.: Biological and pharmacological activity of higher fungi: 20-year retrospective analysis. Cryptogam. Mycol. **27**, 311–333 (2006)

24. Ahmad, M.F., Ahmad, F.A., Azad, Z., Ahmad, A., Alam, M.I., Ansari, J.A., Panda, B.P.: Edible mushrooms as health promoting agent. Adv. Sci. Focus **1**, 189–196 (2013)
25. Bashir, A., Vaida, N., Dar, M.A.: Medicinal importance of mushrooms—a review. Int. J. Adv. Res. **2**, 1–4 (2014)
26. Chang, S.T., Wasser, S.P.: The role of culinary-medicinal mushrooms on human welfare with a pyramid model for human health. Int. J. Med. Mushrooms **14**, 95–134 (2012)
27. Liu, J.: Biologically active substances from mushrooms in Yunnan, China. Heterocycles **57**, 157–167 (2002)
28. Zhang, Y., Mills, G.L., Nair, M.G.: Cyclooxygenase inhibitory and antioxidant compounds from the mycelia of the edible mushroom *Grifola frondosa*. J. Agric. Food Chem. **50**, 7581–7585 (2002)
29. Choi, M.H., Han, H.K., Lee, Y.J., Jo, H.G., Shin, H.J.: In vitro anti-cancer activity of hydrophobic fractions of *Sparassis latifolia* extract using AGS, A529, and HepG2 cell lines. J. Mushroom **12**, 304–310 (2014)

Part VII
Readiness to Cure

Dual-Targeted Therapy and Molecular Imaging with Radiolabeled Nanoparticles

Blanca Ocampo-García, Brenda Gibbens-Bandala, Enrique Morales-Avila, Laura Melendez-Alafort, Menka Khoobchandani, Maydelyd Trujillo-Nolasco and Kattesh V. Katti

Abstract Radiolabeled targeted nanoparticles have been extensively studied for medical applications. Their multifunctionality and multivalency (among other properties) make them suitable candidates to target different diseases by means of pharmacophore groups for molecular, cellular, and/or tissue targeting. They have been used for molecular imaging and as drug delivery systems to improve drug efficacy and decrease side effects by passive accumulation of drugs in healthy tissues. Metallic nanoparticles can be radiolabeled or be radioactive themselves in order to deposit a large amount of energy into malignant cells, which produces irreversible damage. Because of their high surface area, these can be functionalized with small molecules and biomacromolecules for targeted radiotherapy. Moreover, their quantum size effect and resulting properties recently proved to produce hyperthermia. Polymeric nanoparticles are also acquiring importance in molecular imaging as diagnostic and therapeutic agents, due to their biocompatibility, biodegradability, and pharmacokinetic advantages, including the ability for controlled drug release or targeted radiotherapy. Both metallic and polymeric nanoparticles have been proposed as new, smart, pharmaceutical devices to produce dual-targeted therapy and molecular imaging. In this chapter, we will discuss the development and potential medical applications of radiolabeled metallic and polymeric nanoparticles as intelligent targeted systems.

B. Ocampo-García (✉) · B. Gibbens-Bandala · M. Trujillo-Nolasco
Departamento de Materiales Radiactivos, Instituto Nacional de Investigaciones Nucleares, Carretera México-Toluca S/N, Ocoyoacac 52750, Estado de México, Mexico
e-mail: ocampo_be@yahoo.com.mx; blanca.ocampo@inin.gob

B. Gibbens-Bandala · E. Morales-Avila · M. Trujillo-Nolasco
Facultad de Química, Universidad Autónoma Del Estado de México, Paseo Tollocan S/N, Toluca 50180, Estado de México, Mexico

L. Melendez-Alafort
Instituto Oncologico Veneto IOV-IRCCS, via Gattamelata 64, 35138 Padua, Italy

M. Khoobchandani · K. V. Katti
Department of Radiology, Institute of Green Nanotechnology, University of Missouri Cancer Nanotechnology Platform, University of Missouri, One Hospital Drive, Columbia, MO 65212, USA

M. Khoobchandani and A. Saxena (eds.), *Biotechnology Products in Everyday Life*, EcoProduction, https://doi.org/10.1007/978-3-319-92399-4_14

Keywords Radiolabeled nanoparticle · SPECT imaging · PET imaging
Cerenkov imaging · Polymeric nanoparticle · Inorganic nanoparticle
Nuclear imaging · Targeting · Drug release

1 Introduction

Nanoparticles (NPs) play an important role in life science research. Particularly, nanosystems that enable molecular imaging of biological processes and therapeutic applications have been successfully developed based on their capacity to produce multifunctional and multivalent effects.

In the medical field, these nanosensors can be designed to target pharmacophore groups in select cells in order to produce a desired in vitro or in vivo effect. NPs have been purposed as in vitro approaches for diagnosis, in vivo molecular imaging or targeted delivery, and in vivo tissue engineering [1].

For human applications, only a few nanosystems or nanoplatforms have been approved for use in patients. Several shortcomings, mainly safety issues, have challenged their translation to clinical applications. Among them, their inherent toxicity produced by the accumulation in the reticuloendothelial system (RES) originated by their relatively slow hepatic uptake and biliary excretion continues to hamper their widespread use in vivo [2].

Molecular imaging for clinical use refers to the implementation of imaging techniques with highly sensitive and specific recognition with an additional high degree of spatial resolution. Modalities based on nuclear techniques combined with nuclear resonance imaging (NMR) promise a high degree of efficiency, increasing the diagnostic accuracy. Additionally, radiolabeled techniques extend the possibility for their therapeutic application using proper surface architecture to guarantee selectivity by molecular targeting.

Nuclear imaging modalities include single-photon emission computed tomography (SPECT), positron emission tomography (PET), and more recently, Cerenkov luminescence (CL) has also gained prominence. By using these imaging modalities, several approaches based on nanoparticles are being currently studied.

In this chapter, we discuss the development and potential medical applications of radiolabeled metallic and polymeric nanoparticles as intelligent targeted systems for applications in molecular imaging and therapy.

2 Multitargeting Receptors

In several diseases, including cancer, the heterogeneity of overexpressed receptors is well known. This fact opens the possibilities to target concurrently multiple receptors in vivo, in order to improve the detection sensitivity. In general, there are three different strategies for multireceptor targeting [3]:

(a) Heteromultivalent ligands, which allow simultaneous binding to different receptors.
(b) The co-injection of multiple radiotracers.
(c) The sequential injection of different imaging agents.

NPs have emerged as heteromultivalent and multifunctional systems that enable targeting of more than one receptor site to enable both therapy and imaging [4]. In the medical field, multifunctional nanoparticles that combine therapeutic molecules, molecular targeting, and diagnostic imaging abilities and exhibit appropriate features for in vivo use can improve the efficacy of cancer therapy and disease diagnosis. Although most of the nanosystems studied have demonstrated a wide variety of properties, multifunctional and multivalent nanoapproaches that simultaneously exhibit all needed functionalities for human applications are currently limited. Essential functionalities for multifunctional nanocarriers comprise:

- In vivo stability before attainment the target sites.
- Long circulation time in the bloodstream.
- Sensitivity to local stimuli to produce controlled release (temperature and/or pH).
- High drug loading content.
- Ability to specifically accumulate in the target sites.
- Ability to effect the intracellular drug uptake behavior.
- Capability to monitor disease advancement.

Targeting molecules that are capable of attaching onto the surface of nanoparticles include peptides, aptamers, small molecules, and antibodies [5].

3 Nuclear Imaging Based on Nanoparticles

A wide number of nanoplatforms have the capability for chemical conjugation to a chelator or to be radiolabeled on a chelator-free way. The selection of radionuclide to be attached or adsorbed onto the nanoparticle surface depends on the usefulness of that nanoplatform. Some features to be considered include the emission mode, emitted energies, and physical half-life. Gamma-emitting radionuclides with a short half-life are preferred for imaging, and beta-particle emitters are chosen for therapeutic purposes. Radionuclides that enable both imaging and therapeutic capabilities are called "theranostic" (^{177}Lu, ^{90}Y or ^{198}Au). In nuclear imaging, gamma radiation emitted by different radionuclides in diverse decay modes allows the acquisition of images (Fig. 1).

Positron emission tomography (PET) allows measurement of physiologic processes. This imaging can be performed by the annihilation of a positron emitted from a radionuclide with an electron. In this phenomenon, two 511 keV photons are emitted at 180°, which pass through the body and are detected by a ring of detectors around the subject. Radiolabeled NPs can be imaged in a quantitative mode to define the tumor uptake provided by the radiolabeled NPs [5]. Among PET radionuclides,

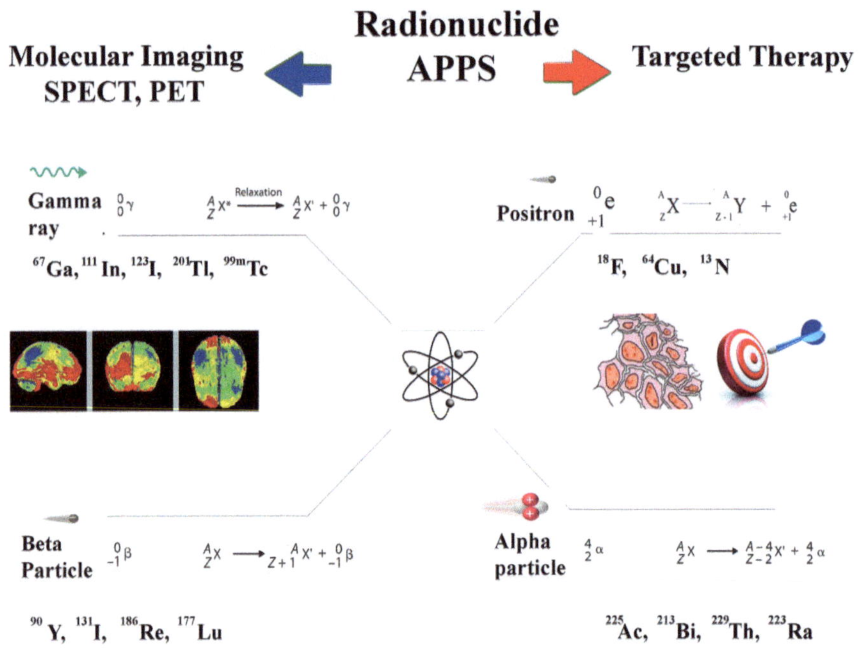

Fig. 1 Decay modes for nuclear imaging and therapy

^{68}Ga ($t_{1/2} = 68$ min) has a $t_{1/2}$ necessary to produce in vivo imaging of NPs and has been used for both PET and Cerenkov imaging. The ^{18}F ($t_{1/2} = 110$ m) radionuclide is also employed to label NP. For longer in vivo PET imaging, ^{64}Cu ($t_{1/2} = 12.7$ h), ^{89}Zr ($t_{1/2} = 78.4$ h), and ^{124}I ($t_{1/2} = 4.18$ d) are commonly used [6].

In single-photon emission computed tomography (SPECT), the emission of a gamma photon originated by the nucleus produces regional foci of the in vivo distribution. Collimators are used to detect a specific range of photon energies. Usually, two opposite detectors are used to obtain images through rotating from multiple angles around the field of view, which are then reconstructed to render three-dimensional images [7].

Cerenkov imaging is produced by the visible light wavelength produced by a charged particle traveling through a dielectric medium faster than the speed of light in that medium [8]. Radionuclides ^{177}Lu and ^{90}Y are used as imaging and therapeutic agents. Moreover, other radionuclides can be used in combination with NPs to obtain a theranostic agent [5].

4 Radiolabeling of Nanoparticles

In general, there are four main strategies to radiolabel nanoparticles:

- Bifunctional approach using chelators to coordinate radionuclides, chemically bound to the nanoparticle surface.
- A chelator-free radiolabeling method whereby radionuclides can be directly bound to the nanoparticle surface.
- Direct bombardment with neutrons or protons to produce some radioactive atoms in the nanoparticle.
- Radionuclides can be embedded into the nanoparticle.

In nuclear imaging, several chelators have been used to bind radionuclides onto the nanoparticle surface. NOTA (1,4,7-triazacyclononane-1,4,7-triyltriacetic acid), DOTA(1,4,7,10-tetra-azacyclododecane-1,4,7,10-tetrayltetra-acetic acid), *p*-SCN-Bn-DOTA (S-2-(4-isothiocyanatobenzyl)-1,4,7,10-tetraazacyclododecane tetraacetic acid), and diethylenenetriaminepentaacetic acid (DTPA) are bifunctional complexing agents which bind radioactive atoms to a target biomolecule.

The conditions of radiolabeling include reaction temperature, incubation time, and biomolecule degradation control, making some modifications on affinity, binding properties, and consequent pharmacokinetic behavior, possible.

5 Radiolabeled Metallic Nanoparticles

Stable inorganic NPs have been prepared using different approaches. Mainly, NPs for biomedical applications are based on nanocolloids formed by heavy metal suspensions stabilized with a coating process using polymers, proteins, or polysaccharides. The coating provides NPs colloidal biocompatibility, stability, and increases their circulation time reducing their uptake by the RES. Nanoparticle biodistribution generally depends on their coating but also on their size. Small NPs with a mean diameter less than 10 nm undergo fast renal filtration; unlike NPs with a diameter greater than 200 nm are quickly removed by the RES system from the bloodstream. Therefore, NPs with diameters between 10 and 100 nm have achieved a higher accumulation at the target site, as a result of their longer circulation times. Furthermore to minimize the opsonization and clearance processes, generally NPs can be coated with some polymers as poly(ethylene glycol) (PEG) [9].

The most investigated inorganic NPs are gold NPs (AuNPs) and iron oxide NPs (IONPs), but there are also some reports of other noble metals such as silver and copper. Nanoparticles offer two key advantages as targeted agents: Firstly, nanoparticle geometry consists of a core, typically with thousands of detectable atoms such as iron and gold; secondly, they can be coated with targeting peptides, antibodies, or any molecules with biological activity. In addition, they have a large surface area, which is ideal not only for efficient modification but also can incorporate various functional moieties on the surface to produce systems with multiple receptor targeting at the same time. This produces multivalent effects caused by multiple simultaneous interactions between the surface of the nanoparticle and that of the cell [10–12].

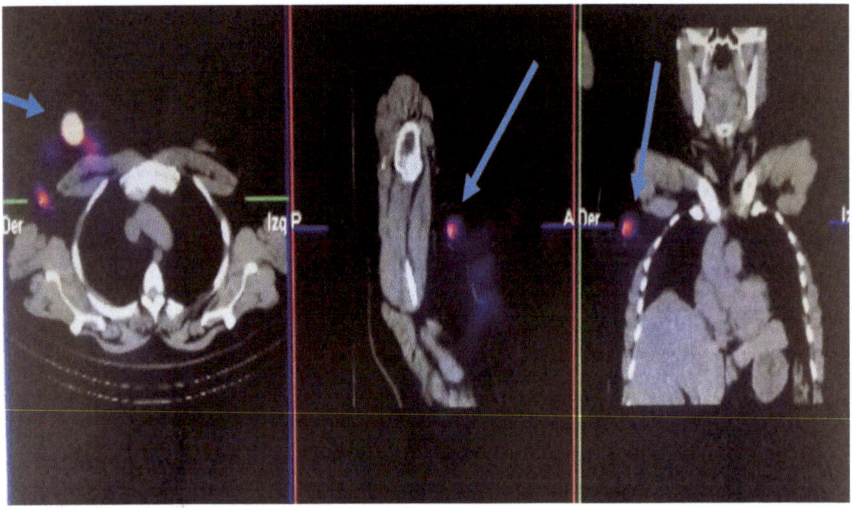

Fig. 2 Sentinel node localization by SPECT/CT in patient with breast cancer (24 h ^{99m}Tc-AuNP-mannose post-administration). Image courtesy of National Cancer Institute (Mexico)

5.1 Radiolabeled Gold Nanoparticles (AuNPs)

Gold nanoparticles continue to show great potential for clinical applications. In recent years, important breakthroughs have been made in the development of gold radiolabeled nanoparticles, which can be used as novel diagnostic tools in multi-modality imaging systems. Some multimeric systems of AuNPs radiolabeled with ^{99m}Tc have been reported as suitable target-specific drugs for molecular imaging of tumors and sentinel lymph node detection [13, 14]. Moreover, ^{177}Lu-AuNPs conjugated to different targeting peptides have been proposed as theranostic radiopharmaceuticals [15–18]. Recently, Ferro-Flores et al. (2017) developed an antiangiogenic cancer-specific dual-targeting ^{177}Lu-Au-nanoradiopharmaceutical based on AuNPs. The nuclear localization sequence (NLS)-Arg-Gly-Asp peptide and an aptamer (HS-pentyl-pegaptanib) to target both the α(v)β(3) integrin and the vascular endothelial growth factor (VEGF) overexpressed in the tumor neovasculature has been demonstrated. The nanosystem showed properties of angiogenesis inhibition [19].

Only few examples of AuNPs have demonstrated successful clinical applications, but several approaches are still in clinical trials [20]. To localize sentinel lymph node in breast cancer patients (Fig. 2), ^{99m}Tc-AuNP-mannose radiopharmaceutical showed a suitable response by using 1-day or 2-day conventional protocols [21].

Plasmonic Gold Nanoparticles for Photothermal Therapy

NPs of noble metals have a broad absorption band in the visible spectrum, due to plasmon resonance. Metallic nanomaterials display strong absorption in the near-infrared (NIR) region (700–1100 nm); mostly, it converts optical energy into thermal energy. This effect is called photothermal ablation and can be used to destroy tumor cells by local heating. Photothermal ablation therapy has gained increasing attention because of its minimally invasive approach for cancer treatment. Several specific targeting approaches for photothermal heating with gold nanoparticles have been widely reported [22–26]. The plasmon resonance for common gold 20-nm nanosphere is 520 nm and redshift in NIR region from 800–1200 nm. Further, these materials can be conjugated with specific targeting molecules for better efficacy [27, 28].

In order to elucidate the temperature increase necessary to produce cell death by photothermal therapy when AuNPs in different tissues are irradiated with a Nd-YAG laser (532 nm), the optical properties (coefficient of extinction, absorption, and scattering) were calculated [29].

Recently, a new generation of nanosystems, which enables more than one pathway of producing therapy and imaging, has emerged. As expected, these systems have demonstrated clear advantages to those of individual therapy approaches. The photothermal and radiotherapeutic potential of the ^{177}Lu-dendrimer conjugated toward folate and bombesin conjugated with gold nanoparticles in the dendritic cavity (^{177}Lu-DenAuNP-folate-bombesin) was demonstrated. The intense NIR fluorescence emitted at 825 nm from the conjugate inside breast cancer cells corroborated the effectiveness of ^{177}Lu-DenAuNP-folate-bombesin for optical imaging [30]. Moreover, the synergistic interaction in a breast cancer model between heat produced by photoconversion and cytotoxicity with doxorubicin was demonstrated by small AuNPs (less than 20 nm) when irradiated by laser [31].

Intrinsically Radiolabeled Gold Nanoparticles

Au-198 can be used for tumor therapy because of its half-life of 2.7 days and higher energy β-emission ($\beta_{\mathrm{max}} = 0.96$ meV) which enables it to penetrate up to 11 mm in tissue to produce therapeutic cross-fire effect on the tumor cells with a minimal radiation dose to the normal tissue surrounded. In vitro/in vivo biocompatible of Au-198 compounds was produced and evaluated the first time by Katti et al. [32, 33]. They prepared trigonal and tetrahedral gold compounds functionalized with biocompatible water-soluble hydroxymethyl phosphine that showed suitable in vivo clearance by renal and hepatobiliary pathways. The present article also demonstrates antitumor properties of these complexes against some types of human cancer established in mice and dogs as lung, breast, non-Hodgkin's lymphoma, prostate, and pancreatic tumors [34]. The promising radiochemical and biodistribution properties of these Au-198 nanoparticles encouraged their stabilization with other polymers as epigallocatechin gallate (EGCG) and gum arabic glycoprotein (GAP) [35]. The utility of amino acid-based phosphines for reducing gold salt into gold nanoparticles has allowed to prepare biocompatible ^{198}AuNPs for theranostic applications [36, 37]. Recently, Axiak-Bechtel et al. used dogs to develop a human-mimicking prostate cancer model. They found that dogs with spontaneous prostatic tumors treated with a single dose of GAP-^{198}AnNP (105 Gy) did not show short-term toxicity.

In addition, the combination of imaging modalities (CT/SPECT) revealed that, following the injection of GAP-^{198}AuNP, the therapeutic agent was mainly localized in the prostate, with some loss in the bladder and urethra [38]. Advantageous Au-198 radiochemical properties also encouraged Chanda et al. to conjugate ^{198}AuNPs with BBN peptides in order to develop a tumor-targeted therapeutic agent. ^{198}AuNP-BBN in vitro and in vivo studies in mice evidenced a high binding affinity (IC_{50}) in microgram ranges and a selective uptake in GRP-receptor-rich organs such as the pancreatic acne in normal mice and the tumors in prostate tumor-bearing mice. The difference in prostate tumor sites between the mice treated with ^{198}AuNP-BBN and the pretreated group demonstrated the realistic clinical potential of these targeted nanoparticles [39]. Radioactive AuNP has been also functionalized with mangiferin, a promising xanthonoid for tumor targeting. The therapeutic efficacy studies of MGF-^{198}AuNPs provided conclusive evidence that the nanosystem has the ability to reduce tumors volume and did not cause any adverse radiotoxicity [40].

5.2 *Radiolabeled Iron Oxide Nanoparticles*

Iron oxide nanoparticles (IONPs) are one of the most widely studied NPs for imaging applications over the past two decades, because they can enable contrast in MRI through the production of localized magnetic field inhomogeneities. IONP size can range from several nanometers to microns and can be classified according to their hydrodynamic diameter into: ultra-small paramagnetic iron oxide (USPIO) NPs below 50 nm; superparamagnetic iron oxide (SPIO) from 60 to 250 nm; and micron-sized iron oxide particles (MPIO) from 1 to 8 μm [41].

Iron oxide NPs are produced by physical, chemical, and biological methods. However, chemical synthesis methods based on the co-precipitation of Fe^{2+} and Fe^{3+} aqueous salt solutions by addition of a base are the most commonly used due to their low production cost and the high yield. In addition, using these methods is easy to control the size, composition, and the shape of the NPs produced, changing some factors as the Fe^{2+} and Fe^{3+} ratio, the type of salts used (e.g., chlorides, sulfates, nitrates, or perchlorates), pH, and the ionic strength. Other chemical methods such as precipitation (solgel and gas/aerosol gel preparation) have been developed [42].

Functionalized IONPs provide the opportunity to develop tumor-specific thermal therapy for metastatic cancer when inductively heated by an externally applied, alternating magnetic field. To be used as targeted agents, however, IONP cores have to be first stabilized with an adsorbed layer of a biocompatible polymer (such as dextran, chitosan, or polymethacrylate) and then conjugate the target-specific molecules such as antibodies, proteins, peptides [9].

IONPs have been radiolabeled with SPECT and PET imaging isotopes using direct, indirect labeling, and doping approaches, in order to develop multimodality imaging agents for dual PET/MR or SPECT/MRI, as well as trimodality imaging (MR/NIR/PET or SPECT) when conjugated to fluorescent near-infrared optical agents [41].

Several dual-mode imaging probes for PET/MRI using ^{64}Cu have been reported. Jarrett et al. designed iron oxide nanoparticles coupled with ^{64}Cu for the diagnosis of vascular inflammation. Glaus et al. reported ^{64}Cu-DOTA-PEG-IONPs showed strong MR and PET signals and stability in mouse serum. ^{64}Cu-bis (dithiocarbamate-bisphosphonate) conjugated to IONPs has also been reported for in vivo lymphatic imaging. To image $\alpha_v\beta_3$ expression, multifunctional ^{64}Cu-labeled IONPs conjugated to the RGD (Arg-Gly-Asp) peptide have been developed and demonstrated a specific glioblastoma tumor-targeting capability by PET/MRI dual imaging [43]. Xie et al. dual-labeled IONPs encapsulated into human serum albumin matrices with ^{64}Cu-DOTA and Cy5.5, and using in vivo PET/NIR fluorescence/MRI trimodality imaging in a subcutaneous U87MG xenograft mouse model, demonstrated a huge accumulation in lesions, a high extravasation rate, and low uptake by macrophages in tumor microenvironment [44].

Both Ga-67 and Ga-68 have been used for multimodal imaging. Jalilian et al. reported a ^{67}Ga-labeled IONP-folate system with adequate cell membrane permeability and paramagnetic properties for thermotherapy. This system showed excellent stability at room temperature, low liver uptake, and high blood circulation after 24 h. Stelter et al. covalently bonded the transfection agent HIV-1 Tat, the fluorescent dye fluorescein isothiocyanate, and ^{68}Ga to IONPs, and demonstrated that the radionanoconjugate can be applied to efficient cell labeling, subsequent multimodal molecular imaging, and possible thermoablative therapy [44].

Madru et al. prepared ^{99m}Tc-labeled IONPs for the SPECT/MRI imaging multimodality of sentinel lymph nodes. The labeling was carried out through polyethylene glycol coated over the solid iron oxide core. SPECT/MRI imaging confirmed its potential applications in the diagnosis of breast cancer and malignant melanoma, due to the accumulation of ^{99m}Tc-IONPs in animal lymph nodes. Shanehsazzadeh et al. evaluated the biodistribution in mice of dextran-coated IONPs labeled with ^{99m}Tc and found high uptake in the reticuloendothelial system.

In order to study the concomitant efficacy of heating injected magnetic nanoparticles, ^{111}In-labeled ChL6 was conjugated to carboxylated polyethylene glycol (PEG) in different-sized, dextran-coated IONPs, with one to two ChL6 antibodies per nanoparticle. Using athymic mice bearing the human breast cancer model, it was observed that heating the nanoparticles with an externally applied AMF caused tumor necrosis in all cases. SPECT imaging showed a tumor uptake of 14% of the injected dose per gram at 48 h. However, although the heating capacity of the large nanoparticles (30 and 100 nm) was several times greater, the tumor-targeting efficacy was significantly less than that of their 20-nm-sized counterparts [43].

Also, beta-particle emitters such as I-131 have been conjugated to NPs for radiotherapy purposes. For example, Liang et al. radiolabeled IONPs with Re-188 using a direct method with a labeling efficiency of 90% and good in vitro stability and

Table 1 Radiolabeled iron oxide nanoparticles for SPECT/PET MR imaging

Nanoparticle	Application	References
^{99m}Tc-IONPs/diethylene triamine pentaacetic acid (DTPA) and 1,4,7-triazacyclononane-triacetic acid (NOTA)	Multimodality contrast agents for sentinel lymph node mapping	[45]
^{99m}Tc-PEG-BP-USPIO/Poly etilenglycol-bisphosphonates (BP)	Visualization of blood vessels and vascular organs with high spatial definition	[46]
^{99m}Tc-USPIO-bevacizumab	Targeted imaging of hepatocellular carcinoma	[47]
^{99m}Tc-PEG-SPIONs	Molecular imaging for sentinel lymph node (SLN)	[48]
^{111}In-antimesothelin antibody (mAbMB)-SPION	Early diagnosis and treatment planning of mesothelin-expressing cancers using SPECT-MR imaging	[49]
^{64}Cu-DTPA-SPION-Fluorochrome	Trimodality reporter for macrophage and inflammatory plaque components	[50]
^{68}Ga/^{111}In-TAT-FITC-aminosilated-SPIONs	Cell labeling for trimodal imaging	[51]
^{64}Cu-PEG-fosfolipid-SPIONs	Dual PET-MRI imaging agent	[52]
^{64}Cu-bifosfonate-dextran-SPIONs	PET-MR dual modality for draining lymph nodes image	[6]
^{64}Cu-DOTA-polyaspartic acid (PASP)-IONPs-RGD	Dual PET and MRI of tumor integrin expression	[53]
Intrinsically radiolabeled [^{59}Fe]-SPIONs	Dual SPECT-MR detection	[54]
$^{67}Ga^{3+}$ and Cu^{2+}-labeled SPIONs	Multimodal PET/SPECT-MRI agent	[55]

demonstrated the ability of ^{188}Re-IONPs to kill liver cancer cells. Cao et al. prepared silica-coated magnetite nanoparticles immobilized with histidine and linked the Re-188 onto their surface, obtaining a labeling yield of 91%. Chen et al. reported the development of ^{131}I-anti-VEGF cross-linked to dextran-coated IONPs and investigated their therapeutic effects in nude mice with induced liver tumors. Tumor growth delay and tumor inhibition were observed. Therefore, their results suggested that the radioimmunotherapy of an intratumoral injection of ^{131}I-anti-VEGF-IONP is effective for the treatment of liver cancer [44] (Table 1).

5.3 Radiolabeled Silver Nanoparticles

Silver NPs (AgNPs) have been used as antimicrobial agents because they can be incorporated into plastics, textiles, and other materials. However, little is known about in vivo trafficking and deposition of AgNPs. Therefore, a few studies have been reported recently to study the accumulation of AgNPs in organs and their toxicological implications. Ichedef et al. [56] reported a synthesis method for radiolabeled silver nanoparticles from proton activation of silver metal powder, enriched in Ag-107, with a 30.7 meV proton beam to produce the γ-emitter Ag-105 g (half-life of 41.29 days). Following the activation, the powder was dissolved in concentrated nitric acid in order to form silver nitrate ($AgNO_3$), which was used to synthesize the ^{105}AgNP. Chrastina and Schnitzer [57] developed a rapid method for the radiolabeling of AgNPs with I-125 in order to track in vivo tissue uptake of silver nanoparticles after systemic administration by SPECT imaging. Biodistribution analysis revealed uptake of the nanoparticles in the liver (24.5% ID/g) and spleen (41.5% ID/g) at 24 h. Similar results were obtained by Ashraf et al. [58] using water-based suspension of bare silver nanoparticles and dextran-coated AgNPs (dextran AgNPs) radiolabeled with Tc-99m. Both ^{99m}Tc-AgNPs and Tc-99m-dextran-AgNPs were mainly accumulated in the liver/spleen region although dextran delayed liver uptake, enhancing the blood retention time. Farrag et al. reported a simple and rapid method for radiolabeling of three types of Ag NPs using I-125, with high labeling yields (>90%), without disturbing the optical properties. After intravenous injection of the radiocompound in normal and solid tumor-bearing mice, they found that ^{125}I-AgNPs was localized in the tumor site for a long period of time [59].

5.4 Radiolabeled Copper Nanoparticles

Synthesis of intrinsically radiolabeled nanoparticles is an emerging concept in cancer theranostics and is expected to play an imperative role in translating nanotechnology research. Therefore, recently Zhou et al. [60] synthesized radioactive ^{64}CuS NPs, in which ^{64}Cu is an integral building block of CuS rather than a chelate to the NPs. These simple to make ^{64}CuS NPs demonstrated to possess excellent stability and to be suitable both for PET imaging and as photothermal coupling agents for photothermal ablation. Furthermore, the ^{64}CuS NPs showed a passive targeting preference over the tumor site and a strong NIR absorption that mediated ablation of U87 tumor cells after either intratumoral or intravenous injection. Based on these results, a viable strategy for a large-scale production (GBq level) of Cu-64 using medium flux research reactors was explored. Biological studies of ^{64}CuS NPs produced with this method on mice-bearing melanoma tumors revealed a significant tumor uptake (4.64 ± 1.71% ID/g) within 4 h post-injection, with good tumor-to-background contrast [61].

A smart nanosystem was developed for tumor-targeting drug delivery and PET/MR imaging. The nanocarrier is based on superparamagnetic iron oxide nanoparticles radiolabeled with ^{64}Cu and demonstrated favorable properties for combined targeted anticancer drug delivery and PET/MRI dual-modality imaging of tumors overexpressing integrin $\alpha_v\beta_3$. The size (hydrodynamic diameter) was 68 ± 2 nm and was pH-sensitive in order to deliver Doxorubicin. The in vivo ^{64}Cu-labeled cRGD-conjugated SPIO nanocarrier uptake was mainly in the tumor and liver, but not in most normal tissues. The system demonstrated good tumor-targeting capability and successful tumor contrast [4].

6 Radiolabeled Polymeric Nanoparticles

Polymeric nanoparticles have been extensively reported as effective carriers to therapeutic pharmaceuticals and are recently emerging as a new class of molecular imaging (MI) agents for detection and treatment of human diseases [62]. An optimal polymeric nanosystem for MI applications possesses the following components: (a) controlled- and sustained-release properties; (b) smaller size (5–250 nm) to facilitate internalization and probing of cells and they do not have rapid renal clearance; (c) their surface is easily modified with molecular signaling and receptor-targeting molecules; (d) payload-carrying capacity delivers high concentrations of imaging agents to desired region; (e) multimodal potential offers visualization in more than one imaging modality; and (f) theranostic capability enables detection and treatment of disease using a single platform [63].

A wide variety of natural or synthetic polymers (chitosan, PLA, PGA, PLGA, PEG, HPMA, and other acrylate derivatives) with outstanding biocompatibility and biodegradability and nanoparticle preparation techniques (nanoparticles obtained by polymerization of a monomer or obtained directly from a preformed polymer) have been described for polymeric nanoparticle production [64]. There is no exclusive polymer for the encapsulation of any therapeutic or imaging agent, and not all molecules can be incorporated in all polymers. Both the physicochemical properties of the polymer and candidate molecule to be incorporated must be considered.

There are two general methods for the generation of polymeric nanosystems for imaging applications. The first is **covalent conjugation** of contrast agents to a polymeric matrix, followed by the formation of nanoparticles by conventional techniques or the conjugation can be in the surface of preformed nanoparticles. The second is the **physical encapsulation** of the contrast agent within polymeric nanoparticles.

The major advantage of grafted polymers is that the contrast agent is covalently bound to the polymer; thus, the burst release does not occur. However, disadvantages include poor loading efficiency and nonhomogeneous distribution of contrast agents within the polymeric matrix. The physical encapsulated systems may have superior advantages over the covalently bound polymer because of high loading efficiency within the polymeric matrix. However, controlling the burst release contrast agents from nanoconjugates within a biological system remains a significant challenge [65].

For nuclear imaging, radionuclides such as ^{11}C, ^{18}F, ^{64}Cu, ^{76}Br, ^{99m}Tc, ^{111}In, and ^{90}Y have been used with a wide range of copolymers to formulate robust nanosized

Table 2 Examples of polymeric nanoparticles used in nuclear imaging

Polymeric system	Agent	Modality	References
^{99m}Tc-PLGA	^{99m}Tc	SPECT	[66]
HPMA-LMA	^{18}F	PET	[67, 68]
Poly(*t*-butyl acrylate), PEG, methyl acrylate, styrene	^{64}Cu	PET	[69]
Poly-glycidyl-methacrylate(poly-2,3-epoxy-propylmethacrylate)	^{68}Ga	PET	[70]
Polyethylene glycol-coated micelles dually labeled	Cy^{7}, ^{111}In	SPECT, NIR, FL	[71]
PVPh	^{124}I	PET	[72]
PAMAM dendrimer-entrapped gold NPs	^{99m}Tc	SPECT/CT	[73]
PEG-b-PPA/DNA micellar nanoparticles	^{111}In	SPECT/CT	[74]
PEGylated dendrimer poly(amidoamine) (PAMAM)-folic acid conjugates	^{99m}Tc	SPECT	[75]
Poly(*N*-vinylimidazole-co-*N*-vinylpyrrolidone)g-poly(D,L-lactide)	^{123}I	SPECT/CT	[76]
Dextran nanoparticles	^{89}Zr	PET/MR	[77]

CT, Computed tomography; *FL*, Fluorescence

delivery systems. Additionally, the fluorescence imaging technique is integrated with polymeric NPs to develop the image-guide drug delivery system to monitor drug pharmacokinetics, intratumoral drug distribution, and drug tumor accumulation in real time.

The versatility of these nanomaterials makes them an attractive platform for developing highly sensitive molecular imaging agents. Table 2 covers the recent use of polymeric nanoparticles as carriers of molecular MI agents.

7 Discussion

Many different types of nanoparticles have been designed and evaluated over the years. Initially, nanosystems were primarily used for therapeutic purposes, that is, for a more efficient delivery of therapeutic drugs to pathologic sites, while reducing their accumulation in potentially endangered healthy tissues. Currently, some therapeutic nanoparticles are used clinically, and more of these so-called nanomedicine formulations are being evaluated in preclinical and clinical trials. However, in recent years, interest for nanoparticles as diagnostic agents has increased.

Radiolabeled nanoparticles developed as a smart multifunctional platform, designed for in vivo imaging and/or therapy have demonstrated high potential to be used in medical applications. The use of noninvasive nuclear imaging modalities such as PET, SPECT, or CL allows for the study of NP biodistribution in animal models, which can provide essential information for clinical translation of radiolabeled nanosystems to human trials. Medical applications of multifunctional nanoparticles could be reached if therapeutic agents could be incorporated in the same platform as molecular targeting agents with diagnostic imaging capabilities.

The potential of using nanoparticles for molecular imaging is compromised because their pharmacokinetic properties are difficult to control. Among the challenges to translate, these nanoapproaches to human application for imaging and/or therapy are the lack of ability to administer them intravenously and their low specific activities are the main shortcomings to deal with. Nanoparticles would either have to be administered by an intratumoral injection or directly deposited on an artery that feeds a target organ to avoid RES in order to achieve therapeutic applications in humans. It is important to note that in order to facilitate the clinical translation of radiolabeled nanoparticles, appropriate dosimetry and toxicological studies should be included.

Acknowledgements This study was supported by the International Atomic Energy Agency (CRP-F22064, Contract No. 18358) and the Universidad Autónoma del Estado de México, through project No. 4288/2017/CI. This research was carried out as part of the activities of the "Laboratorio Nacional de Investigación y Desarrollo de Radiofármacos, CONACyT".

Conflicts of Interest The authors declare no conflict of interest.

References

1. Choi, H., Lee, Y.-S., Hwang, D.W., Lee, D.S.: Translational radionanomedicine: a clinical perspective. Eur. J. Nanomed. **8**(2), 71–84 (2016)
2. Ma, Y., Mou, Q., Zhu, X., Yan, D.: Small molecule nanodrugs for cancer therapy. Mater. Today Chem. **4**, 26–39 (2017)
3. Reubi, J.C., Maecke, H.R.: Approaches to multireceptor targeting: hybrid radioligands, radioligand cocktails, and sequential radioligand applications. J. Nucl. Med. **58**(Supplement 2), 10S–16S (2017)

4. Yang, X., Hong, H., Grailer, J.J., Rowland, I.J., Javadi, A., Hurley, S.A., et al.: cRGD-functionalized, DOX-conjugated, and 64 Cu-labeled superparamagnetic iron oxide nanoparticles for targeted anticancer drug delivery and PET/MR imaging. Biomaterials **32**(17), 4151–4160 (2011)
5. Pratt, E.C., Shaffer, T.M., Grimm, J.: Nanoparticles and radiotracers: advances toward radionanomedicine. Wiley Interdisc. Rev.: Nanomed. Nanobiotechnol. **8**(6), 872–890 (2016)
6. Torres Martin de Rosales, R., Tavaré, R., Glaria, A., Varma, G., Protti, A., Blower, P.J.: ^{99m}Tc-bisphosphonate-iron oxide nanoparticle conjugates for dual-modality biomedical imaging. Bioconjug. Chem. **22**(3), 455–465 (2011)
7. Goel, S., Chen, F., Ehlerding, E.B., Cai, W.: Intrinsically radiolabeled nanoparticles: an emerging paradigm. Small **10**(19), 3825–3830 (2014)
8. Thorek, D.L., Robertson, R., Bacchus, W.A., Hahn, J., Rothberg, J., Beattie, B.J., et al.: Cerenkov imaging-a new modality for molecular imaging. Am. J. Nucl. Med. Mol. Imaging **2**(2), 163 (2012)
9. Abou, D.S., Pickett, J.E., Thorek, D.L.J.: Nuclear molecular imaging with nanoparticles: radiochemistry, applications and translation. Br. J. Radiol. **88**, 2015 (1054). https://doi.org/10.1259/bjr.20150185. PubMed PMID: WOS:000363315800012
10. Montet, X., Funovics, M., Montet-Abou, K., Weissleder, R., Josephson, L.: Multivalent effects of RGD peptides obtained by nanoparticle display. J. Med. Chem. **49**(20), 6087–6093 (2006). https://doi.org/10.1021/jm060515m. PubMed PMID: ISI:000240826200023
11. Montet, X., Montet-Abou, K., Reynolds, F., Weissleder, R., Josephson, L.: Nanoparticle imaging of integrins on tumor cells. Neoplasia **8**(3), 214–222 (2006). https://doi.org/10.1593/neo.05769. PubMed PMID: ISI:000239282800007
12. Khoobchandani, M., Katti, K., Maxwell, A., Fay, W.P., Katti, K.V.: Laminin receptor-avid nanotherapeutic EGCg–AuNPs as a potential alternative therapeutic approach to prevent restenosis. Int. J. Mol. Sci. **17**(3), 316 (2016)
13. Ocampo-García, B., Ferro-Flores, G., Morales-Avila, E., de María, Ramírez F.: Kit for preparation of multimeric receptor-specific ^{99m}Tc-radiopharmaceuticals based on gold nanoparticles. Nucl. Med. Commun. **32**(11), 1095–1104 (2011)
14. Orocio-Rodríguez, E., Ferro-Flores, G., Santos-Cuevas, C.L., Ramírez, FdM, Ocampo-García, B.E., Azorín-Vega, E., et al.: Two novel nanosized radiolabeled analogues of somatostatin for neuroendocrine tumor imaging. J. Nanosci. Nanotechnol. **15**(6), 4159–4169 (2015)
15. Ferro-Flores, G., E Ocampo-García, B., L Santos-Cuevas, C., de Maria Ramirez, F., Azorin-Vega, E., Meléndez-Alafort, L.: Theranostic radiopharmaceuticals based on gold nanoparticles labeled with ^{177}Lu and conjugated to peptides. Curr. Radiopharm. **8**(2), 150–159 (2015)
16. Vilchis-Juárez, A., Ferro-Flores, G., Santos-Cuevas, C., Morales-Avila, E., Ocampo-García, B., Díaz-Nieto, L., et al.: Molecular targeting radiotherapy with cyclo-RGDFK (C) peptides conjugated to ^{177}Lu-labeled gold nanoparticles in tumor-bearing mice. J. Biomed. Nanotechnol. **10**(3), 393–404 (2014)
17. Luna-Gutiérrez, M., Ferro-Flores, G., Ocampo-García, B.E., Santos-Cuevas, C.L., Jiménez-Mancilla, N., León-Rodríguez, D., et al.: A therapeutic system of ^{177}Lu-labeled gold nanoparticles-RGD internalized in breast cancer cells. J. Mex. Chem. Soc. **57**(3), 212–219 (2013)
18. Jiménez-Mancilla, N., Ferro-Flores, G., Santos-Cuevas, C., Ocampo-García, B., Luna-Gutiérrez, M., Azorín-Vega, E., et al.: Multifunctional targeted therapy system based on ^{99m}Tc/^{177}Lu-labeled gold nanoparticles-Tat(49-57)-Lys3-bombesin internalized in nuclei of prostate cancer cells. J. Labelled Compd. Radiopharm. **56**(13), 663–671 (2013)
19. González-Ruíz, A., Ferro-Flores, G., Azorín-Vega, E., Ocampo-García, B., de Maria, Ramírez F., Santos-Cuevas, C., et al.: Synthesis and in vitro evaluation of an antiangiogenic cancer-specific dual-targeting ^{177}Lu-Au-nanoradiopharmaceutical. J. Radioanal. Nucl. Chem. **314**(2), 1337–1345 (2017)
20. Anselmo, A.C., Mitragotri, S.: A review of clinical translation of inorganic nanoparticles. AAPS J. **17**(5), 1041–1054 (2015)

21. Ocampo-García, B.E., Ramírez, FdM, Ferro-Flores, G., De León-Rodríguez, L.M., Santos-Cuevas, C.L., Morales-Avila, E., et al.: ^{99m}Tc-labelled gold nanoparticles capped with HYNIC-peptide/mannose for sentinel lymph node detection. Nucl. Med. Biol. **38**(1), 1–11 (2011)
22. Alberti, C.: From molecular imaging in preclinical/clinical oncology to theranostic applications in targeted tumor therapy. Eur. Rev. Med. Pharmacol. Sci. **16**(14), 1925–33 (2012). PubMed PMID: WOS:000313936400006
23. Pissuwan, D., Valenzuela, S.M., Cortie, M.B.: Therapeutic possibilities of plasmonically heated gold nanoparticles. Trends Biotechnol. **24**(2), 62–67 (2006). https://doi.org/10.1016/j.tibtech.2005.12.004. PubMed PMID: ISI:000235574900004
24. Abadeer, N.S., Murphy, C.J.: Recent progress in cancer thermal therapy using gold nanoparticles. J. Phys. Chem. C **120**(9), 4691–4716 (2016)
25. Fekrazad, R., Naghdi, N., Nokhbatolfoghahaei, H., Bagheri, H.: The combination of laser therapy and metal nanoparticles in cancer treatment originated from epithelial tissues: a literature review. J. Lasers Med. Sci. **7**(2), 62 (2016)
26. Mendoza-Nava, H., Ferro-Flores, G., Ocampo-García, B., Serment-Guerrero, J., Santos-Cuevas, C., Jiménez-Mancilla, N., et al.: Laser heating of gold nanospheres functionalized with octreotide: in vitro effect on HeLa cell viability. Photomed. Laser Surg. **31**(1), 17–22 (2013)
27. Ocampo-Garcia, B.E., Ramirez, F.D., Ferro-Flores, G., De Leon-Rodriguez, L.M., Santos-Cuevas, C.L., Morales-Avila, E., et al.: Tc-99m-labelled gold nanoparticles capped with HYNIC-peptide/mannose for sentinel lymph node detection. Nucl. Med. Biol. **38**(1), 1–11 (2011). https://doi.org/10.1016/j.nucmedbio.2010.07.007. PubMed PMID: ISI:000286574800001
28. Patra, C.R., Bhattacharya, R., Mukhopadhyay, D., Mukherjee, P.: Application of gold nanoparticles for targeted therapy in cancer. J. Biomed. Nanotechnol. **4**(2), 99–132 (2008). https://doi.org/10.1166/jbn.2008.016. PubMed PMID: ISI:000257794200001
29. Carrillo-Cazares, A., Jiménez-Mancilla, N., Luna-Gutiérrez, M., Isaac-Olivé, K., Camacho-López, M.: Study of the optical properties of functionalized gold nanoparticles in different tissues and their correlation with the temperature increase. J. Nanomater. (2017)
30. Mendoza-Nava, H., Ferro-Flores, G., Ramírez, FdM, Ocampo-García, B., Santos-Cuevas, C., Azorín-Vega, E., et al.: Fluorescent, plasmonic, and radiotherapeutic properties of the ^{177}Lu–Dendrimer–AuNP–Folate–Bombesin nanoprobe located inside cancer cells. Mol. Imaging **16**, 1536012117704768 (2017)
31. Mendes, R., Pedrosa, P., Lima, J.C., Fernandes, A.R., Baptista, P.V.: Photothermal enhancement of chemotherapy in breast cancer by visible irradiation of gold nanoparticles. Sci. Rep. **7**(1), 10872 (2017)
32. Berning, D.E., Katti, K.V., Barnes, C.L., Volkert, W.A., Ketring, A.R.: Chemistry in environmentally benign media. 7. Chelating hydroxymethyl-functionalized bisphosphines as building blocks to water-soluble and in-vitro-stable gold(I) complexes. Synthesis, characterization, and x-ray crystal structures of Au{(HOH2C)(2)PC6H4P(CH2OH)(2)}(2) Cl and AU(2){(HOH2C)(2)PCH2CH2P(CH2OH)(2)}(2) Cl-2. Inorg. Chem. **36**(13), 2765–2769 (1997). https://doi.org/10.1021/ic961396b. PubMed PMID: WOS:A1997XF30900013
33. Berning, D.E., Katti, K.V., Volkert, W.A., Higginbotham, C.J., Ketring, A.R.: Au-198-labeled hydroxymethyl phosphines as models for potential therapeutic pharmaceuticals. Nucl. Med. Biol. **25**(6), 577–583 (1998). https://doi.org/10.1016/s0969-8051(98)00023-7. PubMed PMID: WOS:000075024800010
34. Higginbotham, M.L., Henry, C.J., Katti, K.V., Casteel, S.W., Dowling, P.M., Pillarsetty, N.: Preclinical tolerance and pharmacokinetic assessment of MU-Gold, a novel chemotherapeutic agent, in laboratory dogs. Vet. Ther. **4**(1), 76–82 (2003). PubMed PMID: WOS:000220158900008
35. Katti, K.V., Kannan, R., Katti, K., Kattumori, V., Pandrapraganda, R., Rahing, V., et al.: Hybrid gold nanoparticles in molecular imaging and radiotherapy. Czechoslovak J. Phys. **56**, D23–D34 (2006). PubMed PMID: WOS:000206057500003

36. Berning, D.E., Katti, K.V., Barnes, C.L., Volkert, W.A.: Chemical and biomedical motifs of the reactions of hydroxymethylphosphines with amines, amino acids, and model peptides. J. Am. Chem. Soc. **121**(8), 1658–1664 (1999)
37. Kannan, R., Rahing, V., Cutler, C., Pandrapragada, R., Katti, K.K., Kattumuri, V., et al.: Nanocompatible chemistry toward fabrication of target-specific gold nanoparticles. J. Am. Chem. Soc. **128**(35), 11342–11343 (2006)
38. Axiak-Bechtel, A.M., Upendran, A., Lattimer, J.C., Kelsey, J., Cutler, C.S., Selting, K.A., et al.: Gum arabic-coated radioactive gold nanoparticles cause no short-term local or systemic toxicity in the clinically relevant canine model of prostate cancer. Int. J. Nanomed. **9**, 5001–5011 (2014). https://doi.org/10.2147/ijn.s67333. PubMed PMID: WOS:000343863800001
39. Chanda, N., Kattumuri, V., Shukla, R., Zambre, A., Katti, K., Upendran, A., et al.: Bombesin functionalized gold nanoparticles show in vitro and in vivo cancer receptor specificity. Proc. Natl. Acad. Sci. U. S. A **107**(19), 8760–8765 (2010). https://doi.org/10.1073/pnas.1002143107. PubMed PMID: WOS:000277591200049
40. Al-Yasiri, A., Khoobchandani, M., Cutler, C., Watkinson, L., Carmack, T., Smith, C., et al.: Mangiferin functionalized radioactive gold nanoparticles (MGF-198 AuNPs) in prostate tumor therapy: green nanotechnology for production, in vivo tumor retention and evaluation of therapeutic efficacy. Dalton Trans. **46**(42), 14561–14571 (2017)
41. de Barros, A.L.B., Tsourkas, A., Saboury, B., Cardoso, V.N., Alavi, A.: Emerging role of radiolabeled nanoparticles as an effective diagnostic technique. EJNMMI Res. **2** (2012). https://doi.org/10.1186/2191-219x-2-39. PubMed PMID: WOS:000209435700039
42. Kogan, M.J., Olmedo, I., Hosta, L., Guerrero, A.R., Cruz, L.J., Albericio, F.: Peptides and metallic nanoparticles for biomedical applications. Nanomedicine **2**(3), 287–306 (2007). https://doi.org/10.2217/17435889.2.3.287. PubMed PMID: WOS:000248250400008
43. Assadi, M., Afrasiabi, K., Nabipour, I., Seyedabadi, M.: Nanotechnology and nuclear medicine; research and preclinical applications. Hellenic J. Nucl. Med. **14**(2), 149–59 (2011). PubMed PMID: WOS:000292952700009
44. Ferro-Flores, G., Ocampo-Garcia, B.E., Santos-Cuevas, C.L., Morales-Avila, E., Azorin-Vega, E.: Multifunctional radiolabeled nanoparticles for targeted therapy. Curr. Med. Chem. **21**(1), 124–38 (2014). PubMed PMID: WOS:000327778900009
45. Barrefelt, Å.A., Brismar, T.B., Egri, G., Aspelin, P., Olsson, A., Oddo, L., et al.: Multimodality imaging using SPECT/CT and MRI and ligand functionalized ^{99m}Tc-labeled magnetic microbubbles. EJNMMI Res. **3**(1), 12 (2013)
46. Sandiford, L., Phinikaridou, A., Protti, A., Meszaros, L.K., Cui, X., Yan, Y., et al.: Bisphosphonate-anchored PEGylation and radiolabeling of superparamagnetic iron oxide: long-circulating nanoparticles for in vivo multimodal (T1 MRI-SPECT) imaging. ACS Nano. **7**(1), 500–512 (2012)
47. Zhao, Y., Yao, Q., Tan, H., Wu, B., Hu, P., Wu, P., et al.: Design and preliminary assessment of ^{99m}Tc-labeled ultrasmall superparamagnetic iron oxide-conjugated bevacizumab for single photon emission computed tomography/magnetic resonance imaging of hepatocellular carcinoma. J. Radioanal. Nucl. Chem. **299**(3), 1273–1280 (2014)
48. Madru, R., Kjellman, P., Olsson, F., Wingårdh, K., Ingvar, C., Ståhlberg, F., et al.: ^{99m}Tc-labeled superparamagnetic iron oxide nanoparticles for multimodality SPECT/MRI of sentinel lymph nodes. J. Nucl. Med. **53**(3), 459–463 (2012)
49. Misri, R., Meier, D., Yung, A.C., Kozlowski, P., Häfeli, U.O.: Development and evaluation of a dual-modality (MRI/SPECT) molecular imaging bioprobe. Nanomed. Nanotechnol. Biol. Med. **8**(6), 1007–1016 (2012). https://doi.org/10.1016/j.nano.2011.10.013
50. Nahrendorf, M., Zhang, H., Hembrador, S., Panizzi, P., Sosnovik, D.E., Aikawa, E., et al.: Nanoparticle PET-CT imaging of macrophages in inflammatory atherosclerosis. Circulation **117**(3), 379–387 (2008)
51. Stelter, L., Pinkernelle, J.G., Michel, R., Schwartländer, R., Raschzok, N., Morgul, M.H., et al.: Modification of aminosilanized superparamagnetic nanoparticles: feasibility of multimodal detection using 3T MRI, small animal PET, and fluorescence imaging. Mol. Imag. Biol. **12**(1), 25–34 (2010)

52. Glaus, C., Rossin, R., Welch, M.J., Bao, G.: In vivo evaluation of ^{64}Cu-labeled magnetic nanoparticles as a dual-modality PET/MR imaging agent. Bioconjug. Chem. **21**(4), 715–722 (2010)
53. Lee, H.-Y., Li, Z., Chen, K., Hsu, A.R., Xu, C., Xie, J., et al.: PET/MRI dual-modality tumor imaging using arginine-glycine-aspartic (RGD)—Conjugated radiolabeled iron oxide nanoparticles. J. Nucl. Med. **49**(8), 1371–1379 (2008)
54. Hoffman, D., Sun, M., Yang, L., McDonagh, P.R., Corwin, F., Sundaresan, G., et al.: Intrinsically radiolabelled [^{59}Fe]-SPIONs for dual MRI/radionuclide detection. Am. J. Nucl. Med. Mol. Imaging **4**(6), 548 (2014)
55. Pham, T.N., Lengkeek, N.A., Greguric, I., Kim, B.J., Pellegrini, P.A., Bickley, S.A., et al.: Tunable and noncytotoxic PeT/sPecT–MrI multimodality imaging probes using colloidally stable ligand-free superparamagnetic iron oxide nanoparticles. Int. J. Nanomed. **12**, 899 (2017)
56. Ichedef, C., Simonelli, F., Holzwarth, U., Bagaria, J.P., Puntes, V.F., Cotogno, G., et al.: Radiochemical synthesis of Ag-105 g-labelled silver nanoparticles. J. Nanopart. Res. **15**(11) (2013). https://doi.org/10.1007/s11051-013-2073-8. PubMed PMID: WOS:000326322200001
57. Chrastina, A., Schnitzer, J.E.: Iodine-125 radiolabeling of silver nanoparticles for in vivo SPECT imaging. Int. J. Nanomed. **5**, 653–9 (2010). PubMed PMID: WOS:000283715300068
58. Ashraf, A., Sharif, R., Ahmad, M., Masood, M., Shahid, A., Anjum, D.H., et al.: In vivo evaluation of the biodistribution of intravenously administered naked and functionalised silver nanoparticles in rabbit. IET Nanobiotechnol. **9**(6), 368–374 (2015). https://doi.org/10.1049/iet-nbt.2014.0075. PubMed PMID: WOS:000366254300007
59. Farrag, N.S., El-Sabagh, H.A., Al-Mahallawi, A.M., Amin, A.M., AbdEl-Bary, A., Mamdouh, W.: Comparative study on radiolabeling and biodistribution of core-shell silver/polymeric nanoparticles-based theranostics for tumor targeting. Int. J. Pharm. **529**(1–2), 123–133 (2017). https://doi.org/10.1016/j.ijpharm.2017.06.044. PubMed PMID: WOS:000408009200012
60. Zhou, M., Zhang, R., Huang, M.A., Lu, W., Song, S.L., Melancon, M.P., et al.: A chelator-free multifunctional Cu–^{64}CuS nanoparticle platform for simultaneous micro-PET/CT imaging and photothermal ablation therapy. J. Am. Chem. Soc. **132**(43), 15351–15358 (2010). https://doi.org/10.1021/ja106855m. PubMed PMID: WOS:000283621700049
61. Chakravarty, R., Chakraborty, S., Ningthoujam, R.S., Nair, K.V.V., Sharma, K.S., Ballal, A., et al.: Industrial-scale synthesis of intrinsically radiolabeled (CuS)–Cu–64 nanoparticles for use in positron emission tomography (PET) imaging of cancer. Ind. Eng. Chem. Res. **55**(48), 12407–12419 (2016). https://doi.org/10.1021/acs.iecr.6b03405. PubMed PMID: WOS:000389623500012
62. Peng, H., Liu, X., Wang, G., Li, M., Bratlie, K.M., Cochran, E., et al.: Polymeric multifunctional nanomaterials for theranostics. J. Mater. Chem. B **3**(34), 6856–6870 (2015)
63. Choi, K.Y., Liu, G., Lee, S., Chen, X.: Theranostic nanoplatforms for simultaneous cancer imaging and therapy: current approaches and future perspectives. Nanoscale. **4**(2), 330–342 (2012)
64. Reis, C.P., Neufeld, R.J., Ribeiro, A.J., Veiga, F.: Nanoencapsulation I. Methods for preparation of drug-loaded polymeric nanoparticles. Nanomed. Nanotechnol. Biol. Med. **2**(1), 8–21 (2006)
65. Srikar, R., Upendran, A., Kannan, R.: Polymeric nanoparticles for molecular imaging. Wiley Interdisc. Rev.: Nanomed. Nanobiotechnol. **6**(3), 245–267 (2014)
66. Subramanian, S., Dandekar, P., Jain, R., Pandey, U., Samuel, G., Hassan, P.A., et al.: Technetium-99m–labeled poly (dl-lactide-co-glycolide) nanoparticles as an alternative for sentinel lymph node imaging. Cancer Biotherapy Radiopharm. **25**(6), 637–644 (2010)
67. Allmeroth, M., Moderegger, D., Gündel, D., Koynov, K., Buchholz, H.-G., Mohr, K., et al.: HPMA-LMA copolymer drug carriers in oncology: an in vivo PET study to assess the tumor line-specific polymer uptake and body distribution. Biomacromol **14**(9), 3091–3101 (2013)
68. Allmeroth, M., Moderegger, D., Gündel, D., Buchholz, H.-G., Mohr, N., Koynov, K., et al.: PEGylation of HPMA-based block copolymers enhances tumor accumulation in vivo: a quantitative study using radiolabeling and positron emission tomography. J. Controlled Release **172**(1), 77–85 (2013)

69. Sun, X., Rossin, R., Turner, J.L., Becker, M.L., Joralemon, M.J., Welch, M.J., et al.: An assessment of the effects of shell cross-linked nanoparticle size, core composition, and surface PEGylation on in vivo biodistribution. Biomacromol **6**(5), 2541–2554 (2005)
70. Cartier, R., Kaufner, L., Paulke, B., Wüstneck, R., Pietschmann, S., Michel, R., et al.: Latex nanoparticles for multimodal imaging and detection in vivo. Nanotechnology **18**(19), 195102 (2007)
71. Zhang, R., Xiong, C., Huang, M., Zhou, M., Huang, Q., Wen, X., et al.: Peptide-conjugated polymeric micellar nanoparticles for dual SPECT and optical imaging of EphB4 receptors in prostate cancer xenografts. Biomaterials **32**(25), 5872–5879 (2011)
72. Simone, E.A., Zern, B.J., Chacko, A.-M., Mikitsh, J.L., Blankemeyer, E.R., Muro, S., et al.: Endothelial targeting of polymeric nanoparticles stably labeled with the PET imaging radioisotope iodine-124. Biomaterials **33**(21), 5406–5413 (2012)
73. Li, X., Xiong, Z., Xu, X., Luo, Y., Peng, C., Shen, M., et al.: ^{99m}Tc-labeled multifunctional low-generation dendrimer-entrapped gold nanoparticles for targeted SPECT/CT dual-mode imaging of tumors. ACS Appl. Mater. Interfaces **8**(31), 19883–19891 (2016)
74. Patil, R.R., Yu, J., Banerjee, S.R., Ren, Y., Leong, D., Jiang, X., et al.: Probing in vivo trafficking of polymer/DNA micellar nanoparticles using SPECT/CT imaging. Mol. Ther. **19**(9), 1626–1635 (2011)
75. Zhang, Y., Sun, Y., Xu, X., Zhang, X., Zhu, H., Huang, L., et al.: Synthesis, biodistribution, and microsingle photon emission computed tomography (SPECT) imaging study of technetium-99m labeled PEGylated dendrimer poly (amidoamine)(PAMAM)—Folic acid conjugates. J. Med. Chem. **53**(8), 3262–3272 (2010)
76. Lu, P.-L., Chen, Y.-C., Ou, T.-W., Chen, H.-H., Tsai, H.-C., Wen, C.-J., et al.: Multifunctional hollow nanoparticles based on graft-diblock copolymers for doxorubicin delivery. Biomaterials **32**(8), 2213–2221 (2011)
77. Majmudar, M.D., Yoo, J., Keliher, E.J., Truelove, J.J., Iwamoto, Y., Sena, B., et al.: Polymeric nanoparticle pet/mr imaging allows macrophage detection in atherosclerotic plaques novelty and significance. Circ. Res. **112**(5), 755–761 (2013)

Pharmaceutical Biotechnology in Herbal Neuroprotection

Tabassum Zafar, Vinoy K. Shrivastava and Bashirulla Shaik

Abstract Plant-based herbal formulations were the only possible hope for health care management in ancient times. Along with the advancement of social and medical sciences, various drugs came in fashion to treat different diseases and clinical conditions and for a time, the interest in naturally derived compounds is waned. In the current era of biotechnology, where health sciences are at the peak of their advancements, herbal alternatives to synthetic drugs are again gathering tremendous interest. Why are herbal drugs of such interest? Maybe it is good to mention here about the less detrimental side effects of naturally derived compounds. Pharmaceutical biotechnology is a relatively new branch of science, which covers the advancements in pharmaceutical science by the involvement of biotechnological approaches and techniques. Along with the recent advancement of pharmaceutical biotechnology, utilization of compounds derived from medicinal plants for neuroprotection has increased worldwide. Although many synthetic compounds are commercially available nowadays for the patients of neurodegenerative disorders, search for novel herbal options for neuroprotection is still of great significance because of their less non-toxic and long-lasting nature. Without claiming to deal with all features of this topic, this chapter aims to provide a substantial and concise overview of the pharmaceutical biotechnology with a special emphasis in the area of herbal neuroprotection.

Keywords Neuroprotection · Phytochemicals · Herbal biotechnology
Herbal healthcare · Pharmaceutical biotechnology

T. Zafar (✉) · V. K. Shrivastava
Department of Biosciences, Barkatullah University, Bhopal, India
e-mail: tztabassumzafar@gmail.com

B. Shaik
Department of Applied Science, National Institute of Technical Teachers Training and Research, Bhopal, India
e-mail: bshaik@nitttrbpl.ac.in

M. Khoobchandani and A. Saxena (eds.), *Biotechnology Products in Everyday Life*, EcoProduction, https://doi.org/10.1007/978-3-319-92399-4_15

1 Introduction

Medicinal plants and herbs are plants with medical importance for the treatment of different diseases and health conditions. The word "herb" refers to any part of the plant like fruit, seed, stem, bark, flower, leaf, stigma, or a root, as well as a non-woody plant. However earlier, the term "herb" was only applied to non-woody plants, including those that come from trees and shrubs. Medicinal plants are also used widely as food, flavonoid, medicine or perfume and also in certain spiritual activities due to their rich aroma and phytoconstituents. With the advent of civilization, man has learnt many techniques for better survival. Since prehistoric period, man has explored the potential of plants and herbs for curing a wide variety of maladies. Different parts of plants like flower, bark, stigma, and leaves as well as whole plants have been tested for their specific value. Herbs and plants that were studied and explored over a period of time and were designated as the medicinal plant for their specific properties to cure adverse diseases [1].

Recently, the World Health Organization (WHO) estimated that 80 percent of people worldwide relies on herbal medicines for some aspect of their primary health care importance. According to WHO, around 21,000 plant species have the potential for being used for medicinal value. Almost three-quarter of the world's population rely on almost 30 percent of entire plant species for the treatment and cure of primary health concerns. Medicinal plants have great significance in health industry. They have impact on the economy of the developing countries like India in comparison to developed countries. These countries provide two-thirds of the plants used in the modern system of medicine and the healthcare system of rural population depend on indigenous systems of medicine [2].

Biotechnology has promoted tremendous advancement in the field of neurology and has given rise to another new branch of science, called neurobiotechnology, which deals with neuroimaging, neurosurgery, and neuroprotection. This chapter aims to provide a substantial and concise overview of the pharmaceutical biotechnology with a special emphasis in the area of herbal neuroprotection.

2 Neurodegeneration and Neuroprotection

The nervous system is one of the very important body systems. It is responsible to make communication between the body and the brain. Autonomic nervous system and peripheral nervous system work in coordination to perform various vital voluntary as well as involuntary functions. Nervous system transmits signals among various body parts to and from body to perform many vital and voluntary functions. The functions of the nervous system include accomplishment of numerous body processes including vision, sight, smell, hearing, touch, thinking, hunger, thrust, and thermoregulation. There are many disorders that affect the function and coordination of peripheral or central nervous system. These abnormal conditions could cause drastic alterations in humans. The progression of various neurological disorders associated with accident, trauma, stress, and neurodegeneration become more common within the population since last few decades. Performance of somatic, autonomic, sensory, motor, endocrine, and exocrine organs is closely related with the perfect neurological signal transduction. To maintain the health of nervous system is of great importance for vitality. Neuroprotection contributes in management of a healthy nervous system by reducing the nerve damage, nerve degeneration, and autolysis. Neuroprotection could be achieved by restoring antioxidant status, neurotransmitter-mediated pathways, caspases mediated pathways, vitamin mediated pathways, suppressing autoimmune disorders, etc. [3, 4]. To date, there is no synthetic drug available, which is capable of restoring all the nervous system functioning without potential side effects. To overcome such issues, search for potential herbal neuroprotective agents remains a point of interest among researchers.

3 Significance of Herbal Neuroprotectors Over Existing Options

Medicinal plants and herbs have been used since the ancient era to cure different neurological diseases. With all the recent scientific advancements, still, phytochemicals from medicinal plants and herbs are useful in many neurological disorders including dementia, epilepsy, Alzheimer's, Parkinson's, and paralysis. [5–7]. In biological research, the scopolamine-induced animal model of dementia and oxidative stress is widely used as a primary screening test for the determination of anti-Alzheimer effect of unknown plants or drugs. Incorporation to plant-based medical treatment with many other factors also promotes the healing or reduced the severity of neurological damage. Changes in lifestyle is a prompt way to boost immunity against progression and severity of disease. The management of neurological disorders is possible to achieve using dietary restrictions, improved nutrient intake, enhancement of physical and mental activity, obesity management and psychological counselling [8].

Medicinal plants are a very rich source of nutritive compounds along with different antioxidants such as polyphenols, trepenoids, and alkaloids, which are very efficient in removal of reactive oxygen species or free radicals. Free radicals are molecules with unpaired electrons in their outer orbit. Free radical serves many important functions *in vivo*, for example, oxygen radicle participates in many cellular functions. Overproduction of free radicals can initiate a prominent cause of cell damage. Nerve cell damage such as post-ischemic perfusion injury, myocardial infarction, cardiovascular diseases, chronic inflammation, stroke and septic shock, aging, and other degenerative diseases are related with free radicle induced cell damage.

Antioxidant enzymes like such as superoxide dismutase (SOD), catalase (CAT), and glutathione reductase play a vital role in compensating the oxidative stress of cells to make it less vulnerable to cell damage. These enzymes convert harmful hydrogen peroxide into non-toxic molecular oxygen and water.

The antioxidant pool of phytochemicals within the plant body is not only a protector of the plant itself from different kind of the stressors, but also it contributes significantly to the medicinal value of the respective plant [9–11]. Phytochemicals from different sources are used widely for the treatment of countless major and minor diseases. In the treatment of the neurodegenerative diseases, the phytoconstituents of medicinal plants play a crucial role [12].

Herbal medicines from plant-derived chemicals are consistently gaining commercial importance due to many significant properties [13]. Significance statement of medicinal herbs and plants are discussed below:

- Medicinal plants are the only herbal source for disease treatment, which is renewable in terms of economy and environment.
- Cost efficiency, low side effects and convenience of processing are the key features of herbal medicines, which made them popular than allopathic options among both developed and developing countries.
- Conventional herbal medicine systems like homeopathy, Ayurveda, Unani, and Chinese medicine are the vital proof of the efficacy of herbal medicine systems. These conventional systems are still very popular even though we are living in an era of the advanced synthetic drug industry.
- Medicinal plants are a good source of phenols, polyphenols, antioxidants, polyunsaturated fatty acids, short-chain lipid molecules, and fatty acids.
- Purification of phytochemicals is relatively convenient in comparison to the isolation of pure chemical from animal sources as there is no chance of blood-borne infections.
- Plants are the better option in comparison to any animal source or synthetic molecule designed for neurological and other diseases. Plants are the virgin source for medications; phytochemicals are the most popular option for medical treatments for pure vegans.
- Lack of any associated ethical issue also makes the herbal options widely popular.

4 Some Neuroprotective Bioactive Compounds from Plants

Over the last few decades, there has been considerable interest in phytochemical bioactive constituents from herbal medicines due to their long-term medicinal and health-promoting qualities. Some of them have been successfully investigated in animal experiments or clinical trials for potential development into herbal formulations for the treatment of neurological disorders. *Ginkgo biloba*, *Allium cepa*, *Rosa laevigata*, *Fructus chebulae*, *Leonurus heterophyllus*, *Olea europaea*, *Allium sativum*, *Vitis vnifera*, *Vinka minor*, *Mellisa officinalis*, *Narcissus poeticus and Galanthus nivelis*, *Panax ginseng*, *Polygala tenuifolia*, *Crocus sativus*, *Piper methysticum*, *Centella asiatica*, *Ocimum sanctum*, *Lavandula angustifolia*, *and Withania somnifera* are some popular plants and herbs used widely for the treatment of neurological disorders. Plant secondary metabolites and products of secondary metabolism are the key components, which work as bioactive compounds. Asiaticoside, acacetin, magnolol, xyloketal, morin, napthazarin, fucoidan, the aflavin, astragaloside, and tetramethylpyrazine are some bioactive compounds from various plant sources, which are very useful in neurological disorders. These bioactive compounds are the key molecules to treat the underlying mechanism of neurological disorder. These bioactive these chemical components reduce the *in vivo* oxidative stress, enhance motor functions, and boost immunity and memory. These chemicals also affect the level of neurotransmitters such as serotonin, norepinephrine, or dopamine to cure the abnormalities of nervous system [7, 14–17]. Use of *Ginkgo Biloba* improve the effects of memory loss associated with abnormalities in the blood circulation. Phytoconstituents include terpenoids bilobalide, ginkgolides, flavonoids, steroids (sitosterol and stigmasterol), and organic acids (ascorbic, benzoic shikimic, and vanillic acid) are the source molecules behind the unique polyvalent pharmacological action of *Ginkgo Biloba* [18]. Similarly, *Bacopa* includes many active constituents include the alkaloids brahmine and herpestine, saponins d-mannitol and hersaponin and monnierin, betulic acid, stigmastarol, beta-sitosterol repair of damaged neurons by enhancing kinase activity, neuronal synthesis, and restoration of synaptic activity, and ultimately nerve impulse transmission [19, 20].

Several beneficial effects of curcumin on the nervous system (at least ten known neuroprotective actions) have been reported. In an animal model of stroke, curcumin treatment protected neurons against ischemic cell death and ameliorated behavioral deficits [21]. Red grapes contain a high amount of resveratrol, a phytophenol, which exhibits antioxidant potential. It protects neurons in the brain and spinal cord against ischemic injury [22].

5 Discussion

Pharmaceutical formulations have transformed the health sector significantly. In combination with biotechnology, developments in the pharmaceutical industry have been profound and significant. Pharmaceutical biotechnology used to make complex and larger injectable molecules with the help of living cells as primers (such as cells those found in the human body, bacteria cells, yeast cells, and animal or plant cells). Pharmaceutical biotechnology differs from conventional pharmaceutical industry as it involved the synthesis and formulation of complex biological molecules that affect the underlying mechanism of any disease. In some cases where the complete cure is not possible using drugs, still, management of visible symptoms of the disease is possible to achieve using pharmaceutical biotechnology. Management of type 1 diabetes mellitus is a noticeable example of this category, where symptoms of the disease are managed successfully using insulin.

Biotechnology provides a great path for the advancement of pharmaceutical and pharmacogenomics industry by providing opportunities and means to manipulate and modify organisms, including plants at the molecular level. Pharmaceutical biotechnology incorporates recombinant DNA technology for the genetic manipulation of cells to enhance or suppress production of the desired plant-derived compound.

Biotech pharmaceutical products like proteins, antibodies, and recombinant DNA products enlighten new hopes in the area of herbal neuroprotection. Incorporation of pharmaceutical biotechnology on herbal alternative medicine has many advantages over classical allopathic methods to treat neurological disorders [23, 24]. Some of the significant implications of pharmaceutical biotechnology are discussed below:

- **Safety and side effects**: Pharmaceutical biotechnology when paired with herbal and natural sciences become a fine tool to reduce the involvement of synthetic chemical-based harsh drugs. All the chemicals utilized as drugs to treat neurological disorders have many other side effects, while natural compounds are relatively safe and specific to particular clinical symptoms.
- **Specificity**: Involvement of pharmaceutical biotechnology in routine pharmacogenomics and pharmacy-based industry open possibilities to develop more specific medicines according to the genetic makeup and the individual response of patient.
- **Market trends and opportunity**: Involvement of pharmaceutical biotechnology in classical medicine enhances the market opportunities and growth opportunities for producers of medicinal plants and herbs.
- **Bridging between in silico, in vitro, and in vivo**: Pharmaceutical biotechnology creates a realistic approach to achieve accurate and potent results of drug designing and molecular modeling tools. Now, it is possible to synthesize specific medicines in the wet laboratory, which are designed by software according to the specific need.
- **Biochemical engineering**: Biochemical pathways modification in plants for up-regulation of desired phytochemical become possible due to the incorporation of pharmaceutical biotechnology. Development of transgenic plant and herbs to enhance the production of desired chemicals has improved the medicinal value of herbal neuroprotectors.
- **Early diagnosis**: Incorporation of latest technologies such as DNA microarray has increased the probability of early identification and diagnosis of neurological disorders widely.
- **Rapid accessibility**: Although herbal options are quite old concept to retain the health and vitality. Involvement of such latest commercial techniques promotes the rapid accessibility of commercialized form of herbal neuroprotectors worldwide.

6 Conclusions

Pharmaceutical biotechnology is the key to solve mysteries related to classical neurosciences. Pharmaceutical biotechnology is a new hope for health sciences to explore the potential of natural and herbal alternatives for nervous system disorders.

Acknowledgments Authors are thankful to their respective institutions. Ms. Tabassum Zafar acknowledges Department of Biotechnology and the Ministry of Science and Technology, New Delhi, India for providing financial support.

Conflicts of Interest The authors declare no conflict of interest. The funding agency had no role in the design of the study, collection, analyses, interpretation, writing of the manuscript, and the decision to publish the manuscript.

References

1. Wu, P., Zhang, Z., Wang, F., Chen, J.: Natural compounds from traditional medicinal herbs in the treatment of cerebral ischemia/reperfusion injury. Acta Pharmacol. Sin. **31**, 1523–1531 (2010)
2. WHO Neurological Disorders: Public Health Challenges. Available at http://www.who.int/mental_health/neurology/neurodiso/en/. Last Accessed at 20 Nov 2017
3. Le, D.A., Wu, Y., Huang, Z., Matsushita, K., Plesnila, N., Augustinack, J.C., Hyman, B.T., Yuan, J., Kuida, K., Flavell, R.A., Moskowitz, M.A.: Caspase activation and neuroprotection in caspase-3-deficient mice after in vivo cerebral ischemia and *in vitro* oxygen glucose deprivation. Proc. Nat. Acad. Sci. U. S. A. **99**, 15188–15193 (2002)
4. Sultana, R., Perluigi, M., Butterfield, D.A.: Lipid peroxidation triggers neurodegeneration: a redox proteomics view into the alzheimer disease brain. Free Radical Biol. Med. **62**, 157–169 (2013)
5. Uddin, M.S., Nasrullah, M., Hossain, M.S., Rahman, M.M., Sarwar, M.S., Amran, M.S., et al.: Evaluation of nootropic activity of *Persicaria flaccida* on cognitive performance, brain antioxidant markers and acetylcholinesterase activity in rats: implication for the management of alzheimer's disease. Am. J. Psyc. Neurosci. **4**, 26–37 (2016)
6. Da Costa, I.M., Cavalcanti, J.R.L.P., De Queiroz, D.B., De Azevedo, E.P., do Rêgo, A.C.M., Filho, A.I., Parente, P., Botelho, M.A., Guzen, F.P.: Supplementation with herbal extracts to promote behavioral and neuroprotective effects in experimental models of Parkinson's disease: a systematic review. Phytother. Res. **7**, 959–970 (2017). https://doi.org/10.1002/ptr.5813
7. Perry, E., Howes, M.J.: Medicinal plants in dementia therapy: herbal hopes for brain aging? CNS Neurosci. Ther. **17**, 683–698 (2011). https://doi.org/10.1111/j.1755-5949.2010.00202.x
8. Vajeda, F.J.: Neuroprotection and neurodegenerative disease. J. Clin. Neurosci. **9**, 4–8 (2002). https://doi.org/10.1054/jocn.2001.1027
9. Ramawat, K.G. (eds.): Herbal Drugs: Ethnomedicine to Modern Medicine. Springer, Berlin, ISBN 978-3-540-79115-7 (2009)
10. Lobo, V., Patil, A., Phatak, A., Chandra, N.: Free radicals, antioxidants and functional foods: impact on human health. Pharmacogn. Rev. **4**, 118–126 (2010)
11. Harvey, A.L.: Medicines from nature: are natural products still relevant to drug discovery? Trends Pharmacol. Sci. **20**, 196–198 (1999)
12. Orhan, I.E., Loizzo, M.R., Khan, M.T.H.: Therapeutic approaches to neuroprotective activity by complementary and alternative medicines. Evid. Based Complement Alternat. Med. 2012, Article ID 376068, 2 p. http://dx.doi.org/10.1155/2012/376068
13. Brahmachari, G. (Eds.) Discovery and Development of Neuroprotective Agents from Natural Products. Elsevier, Amsterdam, ISBN: 9780128097694 (2018)
14. Ahmad, N., Khan, Z., Hasan, N., Basit, A., Zohrameena, S.: Neuroprotective agents, natural plant herbs & drugs in ischemic stroke: a review. PharmaTutor **5**, 29–36 (2017)
15. More, S.V., Kumar, H., Kang, S.K., Song, K., Lee, K., Choi, D.: Advances in neuroprotective ingredients of medicinal herbs by using cellular and animal models of Parkinson's disease. Evid. Based Complement Alternat. Med. Article ID 957875, 15 p (2013). https://doi.org/10.1155/2013/957875
16. Joshi, H., Parle, M.: Cholinergic basis of memory improving effect of cholinergic basis of memory improving effect of cholinergic basis of memory improving effect of *Ocimum tenuiflorum Linn*. Ind. J. Pharma. Sci. **68**, 364–365 (2006)

17. Kumar, V.M.H., Gupta, Y.K.: Effect of different extracts of *Centella asiatica* on cognition and markers of oxidative stress in rats. J. Ethnopharmacol. **79**, 253–260 (2002)
18. Chandrasekaran, K., Mehrabian, Z., Spinnewyan, B., Drieu, K., Kiskum, G.: Neuroprotective effects of bilobalide, a component of *Ginkgo biloba* extract (EGb 761), in gerbil global brain ischemia. Brain Res. **922**, 282–292 (2001)
19. Anand, T., Naika, M., Swamy, M.S., Khanum, F.: Antioxidant and DNA damage preventive properties of *Bacopa monniera* (L) Wettst. Free Rad. Antioxidants **1**, 89–95 (2011)
20. Singh, H.K., Dhawan, B.N.: Neuropsychopharmacological effects of the Ayurvedic nootropic *Bacopa monniera* Linn. (Brahmi) Indian. J. Pharmacol. **29**, S359–S365 (1997)
21. Capasso, R., Izzo, A.A., Pinto, L., Bifulco, T., Vitobello, C., Mascolo, N.: Phytotherapy and quality of herbal medicines. Fitoterapia **71**, S58–S65 (2000)
22. Huang, S.S., Tsai, M.C., Chih, C.L., Hung, L.M., Tsai, S.K.: Resveratrol reduction of infarct size in Long- Evans rats subjected to focal cerebral ischemia. Life Sci. **69**, 1057–1065 (2001)
23. Lalkovičová, M., Danielisová, V.: Neuroprotection and antioxidants. Neural Regen. Res. **11**(865-87), 4 (2016). https://doi.org/10.4103/1673-5374.184447
24. Misharina, T.A., Fatkullina, L.D., Alinkina, S., Kozachenko, A.I., Nagler, L.G., Medvedeva, I.B., Goloshchapov, A.N., Burlakova, E.B.: Effects of low doses of essential oils on the antioxidant status of the erythrocytes, liver and the brain of mice. Appl. Bioche. Microbiol. **50**, 101–107 (2014)

Considerable Therapeutic Strategies for Treatment of Genetic Defects Govern Neurovascular Disease

Kiranj K. Chaudagar and Abhinav Kanwal

Abstract Vascular malformations are considered as an important substrate for many brain-related diseases. One among them is cerebral cavernous malformation (CCM), which is a mulberry-like blood-filled vascular structure present in the brain. Approximately, 0.5% population has been diagnosed with CCMs worldwide. Although significant progress has been made in exploring the cause of the CCM including gene therapy, understanding of the mutations involved, and the intra- and intercellular pathogenic mechanisms, still no unified theory has been accepted. Keeping this in mind, in this chapter, we are discussing the current understanding about the clinical significance and pathobiology of CCMs followed by various preclinical and clinical developments. Due to severe hemorrhagic disability, asymptomatic nature, and incidental cases, CCM is a significant health burden and is a rare disease. Many United States Food and Drug Administration (USFDA) drugs have been approved for CCMs over the time, but resistance to these therapies is the biggest challenge. Hence, the need for significant and effective therapy is the demand of the time.

keywords Cerebral cavernous malformation · CCM · Therapeutic targets
Drug action · Pharmacotherapy · Gene therapy

1 Clinically Significant Need and Lack of Effective Pharmacotherapy

Biotechnology has revolutionized the pathological understanding of rare and complex diseases. Development of these diseases is usually related to miniscule bio-

K. K. Chaudagar · A. Kanwal (✉)
Department of Medicine, The Pritzker School of Medicine, The University of Chicago, Chicago, IL 60637, USA
e-mail: abhinavkanwal@gmail.com

A. Kanwal
Department of Pharmacology, Manipal College of Pharmaceutical Sciences, Manipal Academy of Higher Education, Manipal, Karnataka 576104, India

M. Khoobchandani and A. Saxena (eds.), *Biotechnology Products in Everyday Life*, EcoProduction, https://doi.org/10.1007/978-3-319-92399-4_16

logical modifications in genes that could be difficult to define without advances in the biotechnology. Biotechnology has provided mechanistic definition, improved diagnostics, and patient care guidelines to these rare diseases. Other than these, it fulfills the need for novel target discovery of gene therapy. The study of brain and cerebrovascular system is usually considered as a complex. Cerebral cavernous malformation (CCM), a rare complex cerebrovascular disease, is a mulberry-like blood-filled vascular structure present in the brain [1]. These caverns genesis, lesion maturation, and related symptomatic phenomena are considered as a CCM disease. An approximately 0.16–0.5% population has been diagnosed with CCM [1, 2]. The population-based annual detection rate is 0.56 per 100,000 per year in adults older than 16 years [1]. Recently, the systemic analyses of CCM history have suggested that the average presentation age of disease is 42.7 years. It is more prevalent in female (52% patients) than male [2]. In contrary to this, the prevalence of CCM is slightly higher in older and younger (50–89 years old and younger than 21 years) male (55.4 and 54%, respectively) than age-matched female [3, 4]. With respect to location, CCM is usually present in lobar (66% cases) and brainstem area (18% cases), rarely found in supratentorial (8% cases) and cerebellar area (8% cases) [2]. CCM patients are suffering from seizure (30%), hemorrhage (26%) and focal neurological deficits (16%, FND) [1, 2]. There are 17% adults and 26% children cases usually diagnosed for CCM, incidentally [3, 4]. However, 20–50% CCM patients are asymptomatic and have been diagnosed incidentally because of wide utilization of brain magnetic resonance imaging (MRI) for neurological diagnosis [1]. Due to severe hemorrhagic disability, asymptomatic nature, and incidental cases, CCM is a significant health burden even it is rare disease.

Genetically, CCM genesis is mainly associated with autosomal dominant mutation in Krev interaction trapped 1 (*KRIT1*), *CCM2,* and programmed cell death protein 10 (*PDCD10*) genes [5]. Approximately, 33% of CCM patients are familial in nature, and among them, CCM1, CCM2, CCM3, and other genes mutated cases are 65, 20, 10, and 5%, respectively [5]. The high prevalence of CCM1 mutation (Q455X) is related to founder effects in Hispanic American population who are located near the southwest region of USA and northern states of Mexico [6, 7]. Ashkenazi Jewish population has CCM2, c.30+5_6delinsTT founder mutation that causes degradation of respective transcript and decreases CCM2 level [8]. The 77.6 kb deletion in exon 2–10 of CCM2 gene is considered as founder effects due to recombination events in US population. CCM2 mutated Italian, French, German, and Swiss populations are different from respective US population [9]. This difference indicates the presence of disease-predisposing elements in CCM gene due to founder effect. Although the prevalence of CCM3 is rarer than CCM1 and CCM2, the severity of disease is high in CCM3 patients than CCM1 and CCM2 [10]. The rate of lesions enrichment with respect to age is high in CCM1 patients than CCM2. Other cases are sporadic in nature and occur due to somatic mutation of these CCM genes [11]. A somatic mutation is found only in endothelial cells but not in surrounding fibrotic tissue, and it is bi-allelic [11]. There is ~20% multiple CCM patients, and many of them are familial [1]. In sporadic patients, multiple lesions formation is governed by the presence of developmental venous anomalies [12]. Thus, CCM progression is usually based on genotyping and anomalous angioarchitecture nature.

Clinically, the CCM disease activity and need of health care are mainly related to lesion hemorrhagic events. Pooled analysis of CCM natural history from seven published studies suggested that the annual hemorrhage rate is 2.5% (1.3–5.1%) per patient-year over 5081.2 patient-years follow-up. The hemorrhage risk is correlated to prior hemorrhage and brainstem location but not with younger age, female sex, lesion size, number of lesion, and developmental venous anomalies. According to prognostic factor-based classification, the 5-year estimated risk of intracerebral hemorrhage (ICH) in patients is 3.8% with non-brainstem CCM presenting without ICH or FND, 8.0% with brainstem CCM presenting without ICH or FND, 18.4% with non-brainstem CCM presenting with ICH or FND, and 30.8% with brainstem CCM presenting with ICH or FND. Here, the hemorrhage risk with ICH CCM patients is also known as a re-hemorrhage risk. Similar to this data, the meta-regression analysis in another study has highlighted annual incidence rate of hemorrhage and re-hemorrhage. An annual incidence rate of hemorrhage is 0.3 and 2.8% per person-year for non-brainstem and brainstem CCM lesion, respectively. The re-hemorrhage annual incidence rate is 6.3 and 32.3% per person-year for non-brainstem and brainstem CCM lesion, respectively. The median time for re-hemorrhage is 10.5 months. The recovery after hemorrhage is 79.5% with minimal disability in 50% patients. For solitary sporadic CCM patients, the annual rates of hemorrhage and re-hemorrhage are 1.03 and 11.95%, respectively. The 5-year risk of re-hemorrhage in these patients is 40.3, and 30% hemorrhagic patients are prone to moderate or severe disability. Due to ~30% risk of re-hemorrhage and at least 20.5% chance of worsening disease activity after each hemorrhagic event, brainstem or prior hemorrhagic CCM patients are considered as critical cases for clinical setting.

Other than hemorrhage, CCM-associated seizure is also a significant symptom for clinical attention. The 5-year risk of first-ever seizure after incidental CCM is 4%, and 5-year risk of epilepsy after first-ever seizure is 94%. This highlights the need of significant therapy for seizure. The 64.7% CCM-related seizure can be controlled by United States Food and Drug Administration (USFDA)-approved drug such as levetiracetam although 26.5% patients are resistant to available anti-epileptic therapy and 8.8% have epilepsy but rare seizure. Herein, CCM management is the only option for anti-epileptic drug-resistant CCM-associated seizure.

From the published evidence, it can be concluded that symptomatic CCM patients are managed by surgery, stereotactic radiosurgery, or propranolol therapy. The systemic review documented higher 5-year risks of hemorrhage (6%) and re-hemorrhage (29.5%) in conservatively managed asymptomatic patients than natural history. Therefore, the surgical decision in CCM patients is based on accessibility of lesions, high risk of bleed, and severity of neurological defects. Particularly, for deep lesion, various surgical approaches are used by skill surgeons to easily access lesion such as anterior transcallosal transchoroidal supracerebellar infratentorial approaches for thalamic lesions and retrosigmoid, suboccipital trans-cerebellomedullary fissure or transnasal transsphenoidal approaches for brainstem lesions. In spite of the application of these approaches, current decision-making knowledge, and advance technologies in surgery, the postoperative outcome is worsened. The 4-year follow-up in microsurgically resected 17 medulla oblongata patients indicated new onset or worsened neurological defects in 6 patients. In surgically resected brainstem CCM patients, the early postoperative morbidity appeared in 29–67% patients, and mortality rate was 1.9%. On long-term follow-up, the 50% resected brainstem lesions were re-bleed, and 0.5% were fatal. In other studies, the 14% brainstem CCM patients had worsened motor symptoms postoperatively even the supportive surgical approaches such as functional MRI, diffusion tensor imaging, and tractography were used. Magnetic resonance thermometry (MRT)-guided stereotactic laser ablation (SLA) rapidly show improvement in disability status of 80% epileptogenic CCM. However, this study is limited due to short follow-up (12–28 months) data, and it needs to provide same beneficial replica on long-term follow-up to prove it worth for broad implications in CCM patients. In CCM-associated seizure patients, preoperative seizure duration and frequency were predictive factors for postsurgical outcome in retrospective study. Apart from these, the patient's decision to undergo surgical management is mainly based on availability of social supports during recovery period. Ultimately, multiple advancements of technologies or even enrichment of knowledge on surgical risk predicting factors are insufficient to resolve CCM symptomatic behavior.

Stereotactic radiosurgery (SRS) is a non-surgical option for eloquent CCM lesion removal. A small single cohort included six CCM patients treated with 13–29 Gy radiation therapy. After 6 months, two patients (33%) re-bleed with increase in VEGF from 26.35 pg/mL (presurgical level) to 65.68 pg/mL (postsurgical level). This means high risk of re-bleed after SRS. These post-irradiated CCM lesions had additional histological features such as EC necrosis and scar formation compared to non-irradiated lesions. This suggests new pathology developed by SRS intervention and limits its application. A SRS outcome meta-analysis in brainstem CCM patients documented 5.61% mortality rate and 11.8% new focal neurological deficits. The relief by SRS management appeared after a long term. The optimal dose of radiation is still a question of debate with respect to safety and effectiveness. Ultimately, SRS is not a good choice for effective CCM management.

Propranolol is approved by USFDA for the treatment of infantile hemangioma, a dermal vascular neoplasm found in children. The benefits of propranolol in infantile hemangioma pertain to vasoconstriction, apoptosis of endothelial cells, loss of angiogenic growth factors, and attenuation of migration–differentiation of endothelial cells. Based on the histopathologic similarity of infantile hemangioma and CCM, propranolol was used off-label for treating CCM. The two propranolol (20 mg three times per day for first year or until tolerable; later 20 mg two times per day up to 2 years)-treated CCM patients (those were unwilling for recommended surgical intervention) highlighted benefits of therapy by decreasing lesion size. Propranolol-related adverse events such as shortness of breath and decrease in exercise tolerance were encountered in one patient and enforced to discontinue treatment before the end of second year. On stopping propranolol administration, this patient again had hemorrhage and lesion growth in 1 year. Therefore, propranolol discontinuation is associated with CCM re-growth and long-term safety of propranolol administration is still a question in CCM patients. Other than these cases, the propranolol treatment checked the CCM hemorrhage in patient with mild von Willebrand disease, no mutation for CCM1–3 gene and non-responsive to tranexamic acid, hemostatic factors, thalidomide, and simvastatin. This case is not appropriate to justify the benefits of propranolol in CCM because of sequential and overlapping treatment of various agents.

According to American Stroke Association/American Heart Association guideline, there is no level A recommendation for any treatment or procedure and very few (eight evidence) are present with Class A recommendation. Collectively, this evidence directs for needs of sustain CCM target-specific pharmacotherapy rather than transient decrease in lesion burden by either nonspecific medications or surgical resection.

2 Pathobiology of CCM Lesions

Structurally, CCM lesion consists of adherent junction-deficient vascular endothelial cells those are without supports of mural cells, basement membrane, and susceptible to transit into mesenchymal cells. These features lead to CCM pertaining pathological changes such as leakiness of vessels (hyperpermeability), non-heme iron deposition, B cell recruitment, and in situ clonal expansion of B cells. All such events are correlated to CCM clinical activity. CCM lesion genesis is mainly based on loss of function mutation in *KRIT1*, *CCM2,* or *PDCD10* gene due to founder or somatic origins. Physiologically, the products of these gene form adaptor complex known as CCM complex which is going to disrupt due to mutation in any one of these genes.

With respect to CCM complex formation nature, KRIT1 is characterized into three distinct structural parts: NPxY/F motif, ankyrin repeats, and C-terminal a band 4.1/ezrin/radixin/moesin (FERM) domain. NPxY/F motif requires for interaction of KRIT1 with phosphotyrosine-binding domain (PTB) domain in CCM2, mitogen-activated protein kinase kinase kinase 3 (MEKK3), and integrin cytoplasmic domain-associated protein 1 (ICAP1/ITGB1BP1). Loss of NPxY/F motif in KRIT1 caused dissociation of CCM2, ICAP1, MEKK3, and activated MEKK3. An activation of this MEKK3–ERK5 axis in brain endothelial cells is a central dogma for CCM lesion genesis. MEKK3 activated at early stage in $Krit1^{ECKO}$ and $Ccm2^{ECKO}$ cells. This led to RhoA activation, Kruppel-like factor 2 (Klf2), Klf4 expression, increased a disintegrin and metalloproteinase with thrombospondin motifs (ADMTS4) proteolytic activity, loss of junctional proteins, and ultimately increased permeability. Following this, familial and sporadic CCM lesions demonstrated increase in MEKK3 activity. Both, $Ccm1^{ECKO}$ and $Ccm2^{ECKO}$ mice, possessed same lesion burden, and MEKK3 heterozygosity induction attenuated lesion development in these mice. Similar to CCM2, CCM2-like protein (CCM2L) exists in complex and is crucial for modulating the function of CCM2–MEKK3 pathway. The expression of CCM2L is tightly regulated by flow. Loss of the flow in CCM lesion decreased expression of CCM2L, activated MEKK3, and worsened CCM disease in rats. In contradiction to these data, $MEKK3^{-/-}$ endothelial cells resulted in hyperpermeability, and multiple organ hemorrhage by increased Rho-associated chain kinase (ROCK) activity mediated loss of vascular endothelial (VE)-cadherin. This was related to CCM complex stabilizing action of MEKK3–CCM2 interaction and role of MEKK3-independent pathway in endothelial hyperpermeability.

CCM2 acts as a linking protein for CCM1 and CCM3 for complex formation. The LD-like motif of CCM2 interacts with focal adhesion targeting (FAT) homology domain of CCM3. CCM2L does not interact with CCM3. $Ccm3^{ECKO}$ cells increased permeability for insulin via extracellular signal-regulated kinase 1/2 (ERK1/2)-dependent cortactin phosphorylation, dissociation of cortical actin from scaffolding protein, zonula occludens-1 (ZO-1). At here, free ZO-1 promoted redistribution of tight junction proteins to cytosol (occludin and claudin-5). These events decreased tight junction (TJ) complex and their protein levels in cytosol without altering respective mRNA expression. Similar to this, cavernoma-lined endothelial cells of sporadic and CCM3 patients decreased TJ complex compared to control.

Other than plasma membrane, CCM3 is found on Golgi apparatus where it regulates exocytosis and plays a pleiotropic role in endothelial junction homeostasis. Endothelial cells are rich in Weibel–Palade body (WPB) that contains angiopoietin 2 (Ang2) and von Willebrand factor (VWF) for secretion. These secretory vesicles possess UNC13 family protein for vesicle docking and priming. The function of UNC13B is inhibited by CCM3 and STK24. $Ccm3^{ECKO}$ mice expressed high amount of Ang2 in brain during maturation of CCM lesions. Secreted Ang2 abolished the membranous localization of VE-cadherin (adherens junction), ZO-1 (tight junction), and connexin43 (endothelial–pericyte junction) by activation of TIE2. It increased number of branch point, lumen diameter, and total number of lumens with loss of pericyte coverage. In support to this, an examination of recurrent human CCM-lined endothelial cells showed high number of WPB under TEM examination. Other than this, permeability control and pericyte coverage inhibited migratory and angiogenic nature of endothelial cells. This pericyte communication was dependent on Notch3 activation through endothelial cells secreted DLL4, and Notch3 functioning impaired in $Ccm1^{ECKO}$ mice. Collectively, CCM1 and CCM2 govern common signaling pathways and CCM3 plays a distinct role in CCM complex. Loss of any protein from CCM complex disturbs its integrity and TJ complex on plasma membrane of EC.

Similar to CCM2, ankyrin repeats and sterile alpha motif containing protein 1B (ANKS1B) contains PTB and interacts with CCM1. In siCCM1-treated endothelial cells, a loss of ANKS1B restored endothelial barrier function without affecting cascades of angiogenesis such as proliferation, migration, and sprout formation. This beneficial action was independent on RhoA inactivation. Consistent with this, Ras-associated protein 1 (Rap1) directly interacts with KRIT1 for vascular homeostasis and prevents hemorrhage. The loss of both, Rap1a and Rap1b, in EC caused hemorrhage in mouse embryo. Endocytosis and endosomal trafficking regulator, sorting nexin 17 (SNX17), had high affinity to NPxY/F motif of KRIT1 in the in vitro experiment. There is still no study to reveal its direct role in CCM lesion pathogenesis. Ultimately, hyperpermeability of endothelial cells is governed by multiple signaling axes; ROCK activation is a prominent downstream cascade of many signaling pathways.

2.1 Endothelial Mesenchymal Transition (EnMT)

Endothelial mesenchymal transition (EnMT) is a crucial phenotypic change for persistence of hyperpermeability other than early-stage permeability modulation. β-catenin plays an important role at the early stage of CCM lesion development by provoking endothelial cells for EnMT whereas TGFβ–pSmad axis leading this EnMT at late stage. More than 70% of EC in human CCM lesion underwent for EnMT. CCM lesions from both sporadic and familial patients have been indicated the role of TGFβ–pSmad axis and β-catenin for EnMT in 90 and 20% endothelial cells, respectively. β-catenin transcriptional activity found in $Ccm3^{ECKO}$ cells of cavernoma and normal brain vessels, whereas TGFβ activity limited to $Ccm3^{ECKO}$ cells of large cavernoma (>150 μM in diameter). This β-catenin activity in $Ccm3^{ECKO}$ cells was independent on canonical pathways (Wnt receptor and its ligands) but governed by Ccm3 loss directed autonomous activation mechanism. Herein, a loss of Ccm3 caused dismantling of adherens junction, loss of VE-cadherin–β-catenin interaction, transportation of β-catenin in the nucleus, and expression of Klf4. Functionally, Klf4 is a zinc finger-containing transcriptional factor that binds on promoter region of EnMT-associated proteins. It strengthened the binding of Smad on promoter region for expression of EnMT proteins and directly enhanced expression of BMP6 for phosphorylation/activation of Smad in $Ccm3^{ECKO}$ cells. At a later stage, β-catenin activity decreased but sustained activity of β-catenin at initial stage was sufficient for phosphorylation of Smad1 and Smad3, an intracellular messenger of TGFβ–BMP axis and provoked late-stage EnMT. Other than β-catenin-dependent axis, $CCM1^{ECKO}$ cells increased expression of Klf4 by activation of MEKK3–MEK5–ERK5 pathway. Disruption of MEKK3 and CCM2 interaction by modifying carboxyl-terminal helical harmonin domain led to MEKK3-mediated Klf4 expression. Human CCM lesions (both familial and sporadic cases) analysis demonstrated the upregulation of Klf4 in endothelial cells surrounding cavernoma. Currently, Klf4 is considered as a biomarker of EnMT in CCM lesions.

Autophagy is a cytosolic bulk cleaning process of cells that digests waste organelles and insoluble proteins. Accumulation of p62/SQSTM1, a marker of autophagy defect, correlated to increase in expression of mTOR, CD44, and ID1, markers of EnMT, in $Krit1^{ECKO}$ cells. $Ccm3^{ECKO}$ cells and human CCM lesions aggregated p62 in endothelial cells. Based on this, defective autophagy is associated with aggresomes accumulation and EnMT in CCM. Additionally, a loss of CCM3 in endothelial cells impaired senescence and autophagy due to loss of cytokine-induced C/EBPβ induction. With respect to EnMT, autophagy is a regulatory cascade for CCM pathobiology.

Other than these, surgically excised CCM lesion examinations have suggested infiltration of various immune cells (B cells, T cells, macrophage, and antigen-presenting DR cells) and their coexistence. These cells were collectively responsible for oligoclonality in IgG by modulating complementary-determining region three sequences in $CD20^+$ B cells. The oligoclonality was lesion specific and not found in blood. In addition, IgM and IgA are detected in few lesions, and IgM predominance over IgA correlated to lesion growth. The number of infiltrated immune cells did not correlate with bleeding or lesion growth. Instead, they represented disease characteristics. $CD20^+$ B cells and $CD3^+$ T cells were high in CCM-associated venous anomaly and solitary lesion, respectively. In young CCM patients, lesions were rich in $CD3^+$ T cells and had rare number of macrophages as compared to adult CCM. This suggests the role of autoimmunity in CCM lesion pathophysiology. However, the molecular cross talk between endothelial barrier integrity and neuroinflammatory sequelae remains to dissect.

Recently, a role of TLR4 stimulation via lipopolysaccharide, endotoxin, and gram-negative bacterial burden in gastrointestinal tract for direct MEKK3 activation has been revealed. In $Krit1^{ECKO}$ and $Ccm2^{ECKO}$ mice, either TLR4 $^{fl/+}$ or $CD14^{-/-}$ genotypic modulation increased lesion volume at least three times, and TLR4 activation accelerated lesion burden more than ten times. CD14 is a TLR4 co-receptor that amplifies intracellular signal transduction pathways of TLR4. VWF and thrombospondin 1 also play a role in CCM pathology. Usually, it is rapidly degraded by proteolytic system after released from endothelial cells. $iCcm2^{KO}$ mice expressed ultra-large VWF multimeric form on cavernous line endothelial cells that speculated for thrombi formation inside some CCM lesions. Thrombospondin 1 (TSP1) is an endogenous angiogenesis inhibitor. A loss of Krit1 suppressed expression of TSP1 in endothelial cells and dismantled tight junctions. The lesion burden in $Krit1^{ECKO}$ mice decreased by re-expression of TSP1 or its anti-angiogenic fragment, 3TSP. In addition, human CCM lesion lined EnMT, and endothelial cells expressed Notch3, and EphB4, respectively. Flk1, VEGF, HIFα, and endoglin were present, respectively, in 71, 41, 48.1, and 63.6% adult CCM lesions endothelial cells. Approximately, 50% of CCM-lined endothelial cells behaved as proliferative by expressing proliferating cell nuclear antigen (PCNA) and nestin. This speculates the role of angiogenic growth factors in CCM development. Till date, the lesion-specific molecular role of only Ang2 is explored.

3 CCM Targets-Based Pharmaceutical Agents

3.1 RhoA–ROCK Signaling Inhibitors

RhoA is found inside the cytosol in bound form with GDP. On activation, guanosine nucleotide exchanger protein replaces GDP with GTP, and RhoA translocates to plasma membrane [13]. Prenylation plays a significant role in translocation of GTP-bound RhoA [13, 14]. After this localization, ROCK binds with RhoA and starts phosphorylation of downstream signaling proteins. The phosphorylation of myosin light chain phosphatase inhibits its own functions and increased the expression of contractile machinery, a phosphorylated myosin light chain [13, 15]. Preclinical and clinical studies demonstrated increase in activation of RhoA and expression of phosphorylated myosin light chain upon inactivation of CCM1, CCM2, and CCM3 in endothelial cells. Therefore, it is a most suitable candidate for drug development. According to RhoA–ROCK signaling mechanism, there are mainly two types of the therapeutic agents for RhoA–ROCK signaling inhibition: (1) RhoA prenylation inhibitors and (2) direct ROCK inhibitor.

Fasudil is approved for the short-term treatment of cerebral vasospasm after subarachnoid hemorrhage as an IV regimen (30 mg infused in 30 min; three times per day for 14 days) in Japan since 1995 but not available in USA [16]. Fasudil is an inactive moiety and metabolized (more than 95% in human) into active hydroxyfasudil, a direct ROCK inhibitor ($Ki = 0.17\ \mu M$) [16, 17]. Placebo-controlled clinical trial of fasudil for cerebral vasospasm management (in 267 Japanese patients) suggested no serious adverse reactions, but infusion decreased 2 mmHg arterial blood pressure for 15 min. Other adverse reactions were intracranial bleeding, hematoma, and reduction in platelet count those were not significantly higher than placebo group [18]. Similar safety profile of fasudil in human has been documented by post-marketing surveillance analysis on 1400 patients and trials on pulmonary arterial hypertension, vasospastic angina, ischemic stroke, and vascular complications of type 2 diabetic patients [16, 17, 19–23]. Preclinical studies have been demonstrated in anti-CCM role of fasudil. Upon chronic administration (>4 months), fasudil (100 mg/kg) was effective to improve survival in $Ccm1^{+/-}Msh2^{-/-}$ mice by decreasing mature lesion burden, iron deposition in lesions, proliferative endothelial cells, recruitment of B cells, inhibiting endothelial ROCK activity and phosphorylated myosin light chain level (pMLC) in EC [24, 25]. These actions were prominent in male mice. Fasudil was ineffective to decrease mature lesion burden in $Ccm2^{+/-}Trp53^{-/-}$ mice. It inhibited hyperproliferating phenotype of $Ccm3^{-/-}$ primary astrocytes (isolated from neonatal mice). This shows that fasudil failed to differentiate the ROCK inhibition selectivity in $Ccm3^{-/-}$ primary astrocytes compared to wild-type primary astrocytes. This provides an importance and need of developing CCM–ROCK selective agents for clinical purpose instead to promote fasudil.

Since three decades statins are well known for lipid lowering, vasculo-beneficial pleiotropic actions, and approved by FDA [26]. Simvastatin (40 mg/kg, >4 months) decreased iron deposition and phosphorylated myosin light chain level in endothelial cells, a marker of ROCK activity, but it did not modulate the lesion burden and survival in $Ccm1^{+/-}Msh2^{-/-}$ mice [25]. Similar to fasudil, simvastatin was ineffective in $Ccm2^{+/-}Trp53^{-/-}$ mice for decreasing lesion burden. Simvastatin and atorvastatin upregulated the expression of CCM2L. Fluvastatin documented most potent CCM-selective RhoA prenylation inhibition in hyperproliferating $Ccm3^{-/-}$ primary astrocytes compared to other short-acting statins (lovastatin, cerivastatin, simvastatin) and long-acting statins (atorvastatin, pitavastatin, pravastatin, mevastatin). In spite of this in vitro experimental proof, it is still indecisive to determine the in vivo CCM–ROCK selective statin.

Each statin is distinct for their pharmacokinetic, toxicological, and pharmacodynamic behavior in humans [27]. Atorvastatin (10–80 mg/day) is considered as a gold standard for prophylaxis of myocardial ischemia or stroke due to pleiotropic actions [28]. A meta-analysis of intensive dose (80 mg/day) atorvastatin or simavastatin therapy suggested benefits by reducing cardiovascular death, myocardial infarction, and stroke through pleiotropic actions. This intensive dose was significantly associated with safety issues such as myalgia-associated withdrawal of patients and liver dysfunction [29]. A recent population-based study suggested no difference between hydrophilic statins and lipophilic statins for intracranial hemorrhage risk in ischemic stroke patients [30]. Another pilot study done on 61 patients reported that hypercholesterolemic patients unable to tolerate other statins including atorvastatin, pravastatin, fluvastatin, simvastatin, lovastatin but rosuvastatin at doses of 5 and 10 mg/d + diet was well tolerated, effective, and had a good safety profile [31]. The study pertaining to effect of simvastatin on permeability in CCM patients has completed, and data are not published yet. Therefore, atorvastatin is a first-choice candidate for CCM development, and later one is rosuvastatin.

Zoledronic acid is a FDA-approved drug for treatment of osteoporosis in postmenopausal and multiple myeloma patients. It acts as a prenylation inhibitor by interacting with geranyl–geranyl pyrophosphate synthase and farnesyl pyrophosphate synthase. It shows synergistic effects in combination with fluvastatin and this combination extended the longevity and decreased lesion burden in endothelial cell-selective Ccm3$^{-/-}$ mice. These drugs had beneficial anti-CCM actions by inhibiting JNK and ERK1/2 but not by KLF2/4. However, there is a lack of in vivo preclinical data to support synergistic interaction between zoledronic acid and fluvastatin. The combination of zoledronic acid (4 mg infusion every 4 weeks) and pravastatin (10 mg/day) or fluvastatin (40 mg/day) or atorvastatin (20 mg/day) has been studied in human to prevent bone metastasis of renal cell carcinoma and bone damage in Hutchinson-Gilford Progeria syndrome [32]. These trials documented safety of using statins in combination with zoledronic acid. This synergism concept will help in clinic to reduce adverse dose of statins with same efficacy.

Due to lack of in vivo CCM selectivity and safety data of statins, it is appropriate to look for new therapeutic agent. ROCK1 and ROCK2 are isoforms of ROCK present in mammals, and they are different in their distribution in tissues [33]. ROCK2 is expressed in heart, brain, and muscle, whereas ROCK1 is found everywhere in tissues except muscle and brain [13, 15]. Because of this difference, ROCK1 plays a role in circulating inflammatory cells processing whereas ROCK2 involves in contraction of vascular smooth muscle [13]. Brain endothelial cells are selectively expressing high amount of ROCK2 [15]. BA-1049 is a novel ROCK2-selective inhibitor and currently under preclinical development for CCM [15]. Ultimately, CCM–ROCK inhibition hypotheses can be satisfied by brain-selective ROCK inhibitor.

3.2 B Cell Depleting Agent

Autoimmunity plays a critical role in the genesis of neurological diseases (multiple sclerosis (MS) and neuromyelitis optica) [34]. In consistent with these diseases, a pathobiological investigation of excised human CCM lesions has strongly supported the involvement of CD20^{+} B cell, oligoclonality, and neuroinflammation in CCM lesions development. In our previous study on both *Ccm3*$^{+/-}$ (mild phenotypic) and

Ccm3$^{+/-}$*Trp53*$^{-/-}$ (strong phenotypic) mice, long-term treatment (12–15 weeks) of anti-mouse BR3 antibody (5 mg/kg/week) significantly decreased number of mature lesion burden, non-heme iron deposition, endothelial ROCK activity without modulating immature lesion burden. These effects were related to depletion of lesion B cell in response to peripheral B cells depletion. Based on this data, B cells are mainly involved in process of lesion maturation.

Various anti-CD20 monoclonal antibodies (rituximab, ibritumomab tiuxetan, tositumomab-I-131, ofatumumab, obinutuzumab, and ocrelizumab) are approved for the treatment of CD20^{+} B cells driven diseases such as MS, rheumatoid arthritis, chronic lymphocytic leukemia, and non-Hodgkin's lymphoma [34, 35]. Among these, rituximab (Rituxan approved in 1997 by USFDA and invented by Genentech), ofatumumab (Arzerra approved in 2009 by USFDA and launched by Novartis), and ocrelizumab (Ocrevus approved in 2017 by USFDA and invented by Genentech) have been tested for management of MS, and only Ocrevus is a USFDA-approved treatment for relapsing-remitting or primary progressive MS [36–40].

As compared to rituximab, which is a chimeric antibody, ofatumumab and ocrelizumab are humanized monoclonal antibodies and favor their regulatory approval with respect to safety concerns so these drugs have been prominently focused for repurposing use in MS [34]. Ocrelizumab is humanized, glycosylated monoclonal antibody that binds with CD20 antigen present on the surface of pre-B cells and mature B cells receptor [34, 36]. Due to this, ocrelizumab facilitates systemic B cells loss by direct induction of apoptosis, complement dependent cytotoxicity (CDC), and antibody-dependent cell-mediated cytotoxicity (ADCC). This systemic depletion of B cells for long term (6 months) ultimately decreased availability of peripheral B cells for local neuroinflammation [36, 41]. Compared to rituximab, ocrelizumab is five times potent for ADCC, three times potent for CDC, and similar in potency for apoptosis of B cells. Both, Rituxan and Ocrevus are invented by Genentech. With respect to long-term legal marketed protection and strong technical background, Genentech has proposed and launched Ocrevus of management of MS. In case of ofatumumab, Phase-III trial data are still pending after announcement of trial initiation in Sep 2016 by Genmab, a product-specific partner of Novartis. In randomized, double-blind, placebo-controlled Phase-II trial, 700 mg infusion of ofatumumab at 24 weeks interval well-tolerated and infusion-related reactions were 100% in ofatumumab 700 mg versus 25% in placebo [37].

Ocrevus is an intravenous injection available as a single vial with strength of 300 mg/10 mL [41]. USFDA recommended OPERA (randomized, double-blind and active controlled Phase-III) trials administration schedule for Ocrevus with considering individual feasibility for hospital visits [41]. Based on OPERA trials, hepatitis B virus infection negative patients eligible for Ocrevus infusion and first dose (600 mg) should be divided into two equal doses administered at an interval of 14 days because of high infusion-related reaction with first dose (35%). Subsequent dose (600 mg) should be given one time at every 24 weeks [38]. Before every infusion, patients should be premedicated with prophylactic methylprednisolone and diphenhydramine to reduce the risk of infusion-related reactions [38]. However, in ORATORIO (randomized, double-blind, and placebo-controlled Phase-III) trial, most of infusions-related adverse reactions were mild in nature and managed by flow rate adjustment. There was no significant difference in patient withdrawal rate between ocrelizumab-treated group and placebo due to adverse reactions [39]. The risk of respiratory infection (59% vs. 14%) and malignancy (2.3% vs. 0.8%) was higher in Ocrevus-treated groups compared to placebo [39]. According to current data on safety profile of ofatumumab and ocrelizumab in multiple sclerosis trials, both of them are prominent for CCM clinical development.

3.3 Vitamin D Analogue

Vitamin D, particularly calcifediol class, acts as a micronutrient and regulates calcium balance in human body through activating vitamin D receptor (VDR), a member of thyroid–steroid nuclear receptor family. Biologically, an active form of vitamin D in human is a 1,25-dihydroxycholecalciferol (calcitriol, active vitamin D3), which is synthesized by kidney from precursor, 25-hydroxycholecalciferol (calcifediol, partially active vitamin D3). Human can get calcidiol by two routes: (1) synthesized by sequential transformation of 7-dehydrocholesterol in skin and liver during sun exposure and (2) available from diets or dietary supplements. Even these two routes, the serum 25-hydroxyvitamin D3 level in 88.1% populations from developed, developing, or underdeveloped countries was lower than recommended level (75 nmol/L) according to meta-analysis which included data of vitamin D3 monitoring trials around the world. Other than calcium homeostasis, various clinical studies suggested an inverse association between vitamin D3 level and severity of neurological disease. Low serum level of 25-hydroxyvitamin D3 exposed persons for risk of dementia, highlighted in longitudinal follow-up Finnish study. The Extended Disability Status Scale (EDSS) more than 4 predicted by plasma vitamin D3 level (below 20 ng/mL) in relapsing-remitting MS patients. In CCM patients, low 25-hydroxyvitamin D3 levels are associated with prognosis or early onset of malignancies related to onset of lesions, multiplicity of lesions and various other hematalogical malignancies. Therefore, 25-hydroxyvitamin D3 is a pleiotropic micronutrient for CCM and other neurological diseases.

Preclinical studies have reinforced hypothesis on pleiotropic beneficial role of vitamin D3 in CCM. In endothelial-specific Ccm2 knockout mice, approximately 50% of lesion burden attenuated by long-term treatment (5 months) with cholecalciferol (25 IU/g). Its acute treatment also halted modulation in structural–functional phenotypes in siCCM2 RNA-treated endothelial cells and abolished vascular leak in endothelial-specific Ccm2 knockout mice. Cholecalciferol, but not its precursor 7-dehydrocholesterol, decreased RhoA expression and myosin light chain activation in *Ccm2* knockdown endothelial cells. The upregulated expression of Klf2 and Klf4 did not attenuate by vitamin D3 treatment in $Ccm1^{ECKO}$ mouse cells. This cholecalciferol and other vitamin D analogues (ergocalciferol, calcifediol, calcitriol, doxercalciferol, calcipotriene, and paricalcitol) are approved by USFDA for treatment of rickets, hypoparathyroidism, hypophosphatemia, plaque psoriasis, and secondary hyperparathyroidism due to chronic kidney disease. Due to 100 times rapid systemic clearance of calcipotriene than calcitriol, it is suitable for topical administration, only. Cholecalciferol is available in combination dosage forms with other nutrition or anti-osteoporotic agent. Other vitamin D analogues (Drisdol (ergocalciferol, 50KIU capsule, US Pharm); Rayaldee (calcifediol, 0.03 mg, extended-release capsule, OPKO); Rocaltrol (calcitriol, 0.25/0.5 μg capsule, Validus Pharm); Hectorol (doxercalciferol, 0.5/1/2.5 μg capsule, Genzyme); and Zemplar (paricalcitol, 1/2 μg capsule, Abbvie)) are available in oral dosage form as a single therapeutic ingredient.

Hypercalcemia and related vascular calcification are major drawbacks of vitamin D analogues clinical use. Preclinical and clinical data suggested vasculo-protective action of paricalcitol, whereas calcitriol did not modulate vascular response. This protection was independent on increase in serum calcium and decrease in parathyroid hormone level. Based on this data, paricalcitol is the first choice among vitamin D analogues for CCM development. There is no data on vasculo-protective potential of other vitamin D analogues (ergocalciferol, calcifediol, and doxercalciferol) treatment. However, they are safe for hypercalcemia risk compared to calcitriol. Ergocalciferol and calcifediol are transformed by kidney into active metabolites, whereas doxercalciferol transformed into active metabolites by liver. Therefore, they are prone to subsequent CCM development. Apart from these, VS-105 (developed by Vidasym, Chicago) is a pleiotropic effect-specific vitamin D receptor agonist. It attenuated ventricular abnormalities and improved endothelium-dependent aortic relaxation without affecting serum calcium level in nephrectomized rats with decreasing parathyroid hormone concentration. Currently, it is under clinical development phase for treatment of osteoporosis, and it would be an appropriate future target for CCM development.

3.4 EnMT Inhibitors

EnMT is a migratory phenotypic modulation in endothelial cells that lead to loss of cell–cell contacts and persistent neuroinflammation [42]. EnMT induction is governed by activation of Notch, Wnt/β-catenin, and TGF-β-signaling pathways [43]. Mesenchymal stem cell markers', CD44 and α-smooth muscle actin, analysis in CCM lesions demonstrated EnMT in pristine EC line. In constituent to this, preclinical evidence suggested that initiation and persistent of EnMT in CCM is regulated by activation of two mechanism i.e β-catenin and TGF-β signaling respectively [43]. Therefore, both the pathways are considered as an important targets for the development of CCM therapeutic agents.

Sulindac sulfide and sulindac sulfone are metabolites of nonsteroidal anti-inflammatory drug, sulindac. Sulindac sulfide is an active metabolite and inhibits cyclooxygenase but sulindac sulfone is not. The short-term treatment (3–9 days) of both sulindac metabolites (30 mg/kg) improved survival and decreased lesion burden in endothelial-specific *Ccm3* knockout mice. These benefits are due to the relocalization of VE-cadherin and recruitment of β-catenin at plasma membrane. Loss of

EnMT driving proteins expression in the endothelial cells also helps in improving the survival. However, due to devoid of COX inhibitory action in sulindac sulfone (exisulind), it is the preferential choice for CCM clinical development by selectively targeting β-catenin pathway. Because of phosphodiesterase V (PDEV) inhibitory and related antineoplastic action, it was under clinical development alone or in combination with other antineoplastic agents for treatment of prostate, lung, and breast cancer. Even exisulind was well tolerated in cancerous patients, it did not improve efficacy and failed for treating prostate, lung, and breast cancer [44–48]. In 2015, it has been received orphan drug designation for CCM from European Medicines Agency (EMA) on request of Firc Institute of Molecular Oncology (IFOM).

LY-364947, LY-2109761, and SB-431542 (TGF-β and pSmad signaling inhibitors, respectively) attenuated lesion burden and restored endothelial–pericyte cells interaction in $Ccm1^{ECKO}$ mice [43]. However, all of these are experimental agents. LY2157299 (galunisertib, invented by Eli-Lilly) and pirfenidone (invented by Intermmune Inc) are TGFβ receptor and synthesis inhibitors, respectively [49, 50]. Pirfenidone (Esbriet, 267 mg capsule/tablet, 801 mg tablet for oral administration up to 2403 mg/day, produced by Genentech) is approved by USFDA for treatment of idiopathic pulmonary fibrosis. In ASCEND trial, 52 weeks treatment of pirfenidone caused significant increase in gastrointestinal (5.4% vs. 1.4%, grade 3 reactions) and skin (1.8% vs. 0.4%, grade 3 reactions) adverse reactions compared to placebo but they were mainly mild or moderate in nature and did not lead to treatment discontinuation [49, 51]. Galunisertib is under clinical development for glioma, hepatocellular carcinoma, and pancreatic cancer [52–55]. In Phase-II recurrent glioblastoma trial, galunisertib (300 mg/day, 14 days on/14 days off intermittent dosing) was well tolerated and improved overall survival compared to lomustine [52]. Therefore, pirfenidone and galunisertib are primary options for CCM clinical development with respect to TGFβ specific therapy.

3.5 TLR4 Inhibitors

TLR4 is an endotoxin recognizing receptor that activates innate immunity at early age of infections. For this physiology, it interacts with cluster differentiation antigen 14 (CD14), lipopolysaccharide (LPS)-binding protein (LBP), and myeloid differentiation protein 2 (MD2), those accelerates TLR4-mediated immune activation. Depending on this, overactivation of TLR4 is critical for multiple organ failure in sepsis by inflammatory events. In addition to sepsis-related inflammation, TLR4 is activated by endogenous ligands such as defensin, heat shock proteins, high mobility group box1 and causes sterile inflammation in absence of endotoxin. It is one possible mechanism for the occurrence of neuropathic pain. In CCM, TLR4 and CD14 expression enhancing polymorphism have been correlated to high lesion burden in Q455X Hispanic founder mutant-carrying patients. TLR4 inhibited CD36-mediated phagocytosis of brain-deposited RBC in microglia. Recent discovery on involvement of endothelial-specific TLR4 in CCM lesion development urges for the TLR4 antagonist in the treatment of CCM.

TLR4-mediated septic inflammation is a sequential and multiprotein containing cascade so various types of TLR4 antagonists have been discovered and screened such as LPS sequestrants, CD14 inhibitors, TLR4 inhibitors, and TLR4 intracellular signaling modulators. None of these TLR4 antagonists are approved for the treatment of sepsis. Eritoran (E5564, invented by Eisai) and TAK-242 (Resatorvid, invented by Takeda Pharmaceuticals) are selective TLR4 antagonists which are in advanced clinical stage. There was no significant change in primary outcome of trials, and eritoran failed to replicate the beneficial data of Phase-II sepsis trial. Therefore, eritoran further development stopped for treatment of sepsis. However, this trial provided the safety data of eritoran to foster other indications. Recently, it is under Phase-II clinical investigation for diabetes. Similar to eritoran, TAK-242 failed for significant improvement in primary outcome of sepsis patients. This trial provided safety signal for TAK-242-related transient dose-dependent methemoglobinemia. TAK-242 had TAK-242 tested at preclinical stage for CCM, and it abolished lesion development in *Krit1*^{ECKO} mice. Therefore, these agents are good choices for preclinical evaluation in CCM experimental models.

Apart from these new agents, the TLR4 inhibitors screening experiments highlighted potential blocking activity of marketed drugs such as naloxone, naltrexone, and ibudilast. The inhibitory activities of naloxone and naltrexone are non-classic and non-competitive on TLR4. Naloxone (Evzio = 0.4 mg or 2 mg/0.4 mL prefilled autoinjector for intramuscular use by Kaleo Ltd; Narcan = 2 mg or 4 mg/0.1 mL nasal spray by Adapt Pharma) and naltrexone (Vivitrol, 380 mg/vials for intramuscular use by Alkermes) are opioid receptor antagonists and approved to treat opioid overdose/dependence or alcohol dependence in USA. Evzio and Narcan are recommended for short-term emergency use, and Vivitrol is extended release for long-term prescribed dosage forms. The common safety signals with Vivitrol use in patients are hepatotoxicity and injection site reactions. The 6 months administration of Vivitrol did not cause any significant change in liver chemistry and hepatic abnormality of alcohol dependence patients. Other than these, the opioid overdose vulnerability and opioid withdrawal precipitation are also naltrexone/naloxone administration associated with adverse risk in opioid prescribed populations. Naloxone and naltrexone are stereoselective for opioid antagonism (means only (+)-forms act as antagonist but not (−)-forms) and non-stereoselective for TLR4 antagonism. Therefore, stereoselective (+)-naloxone/naltrexone are possible choices to develop safe and selective TLR-4 antagonist for CCM according to outcome from future study in CCM experimental models.

Ibudilast is approved in Japan and Korea for treatment of asthma (20 mg/day) and post-stroke syndrome (30 mg/day) more than two decades, but it is not available in USA. Due to phosphodiesterase inhibition-related neuroprotective action, ibudilast is under clinical development for multiple sclerosis in USA. In this Phase-II relapsing-remitting MS trial, 60 mg/day ibudilast administration was safe for 12 months administration and also efficient to decrease disability progression. Means, there will be high probability of USFDA to allow clinical trial of ibudilast in CCM if it will prove benefits in CCM experimental model. Currently, due to lack of preclinical efficacy data on these TLR4 antagonists in CCM, it is required to screen them in respective murine experimental models.

3.6 Thrombospondin 1 Mimetics

In neurological diseases such as subarachnoid and intracerebral hemorrhage, TSP-1 is elevated in plasma but its therapeutic implication is not concluded yet [56–58]. In CCM, the expression of TSR, an anti-angiogenic peptide of TSP-1, abolished dismantling of tight junctions and CCM lesion burden in *Krit1*^{ECKO} mice. This recent preclinical data clearly provide insight for beneficial anti-CCM role of TSP-1. Small molecules such as ABT-510 and ABT-898 (both developed by Abbott) have been discovered for TSP-1 mimetic action [59, 60]. ABT-510 (developed by Abbott) is naturally derived from TSP-1 and possesses strong anti-angiogenic action. Currently, it is under clinical Phase-I/II trial in combination with standard chemoradiation therapy or alone for treatment of glioblastoma and other advanced solid tumors (colorectal, renal, breast, sarcoma, and melanoma) [59, 61–65]. In Phase-II soft tissue sarcoma trial, ABT-510 was achieved more than 100 ng/mL in plasma and remained stable at least more than 3 h/day in all patients after subcutaneous administration of 100 mg ABT-510 two times in a day. This dose schedule was efficient for anti-angiogenic actions of ABT-510 [59]. At same dose, ABT-510 (up to 28 days administration) was well tolerated, and two patients had treatment-associated adverse events. One patient had grade 4 hyponatremia that caused withdrawal of patients, and another patient had grade 3 lymphopenia [59]. Preclinical study of ABT-898 depot preparation highlighted convenience of this new formulation for long-term administration with tolerable safety profile compared to ABT-510 bolus administration [60]. There was a thrombospondin 1 mimetic antibody, CVX-045 (a product generated by CovX technology), under clinical development for advanced solid tumor treatment in 2008 [66, 67]. Later, Pfizer stopped research pipeline on CovX–Body technology due to financial priority. Therefore, it seems promising to evaluate both ABT-510 and ABT-898 for preclinical anti-CCM action.

3.7 Antioxidant Agents

A plenty of preclinical and clinical evidence has been indicating the role of oxidative stress in neurological diseases by modulating vascular barrier functions and hemorrhage activity [68–70]. Oxidative stress is indicated by presence of reactive oxygen species (ROS) which are generated by mitochondria metabolic system. CYP activity is responsible for ROS production, and MMP activity is modified by ROS. The ROS generation by oxidative stress plays a crucial role in CCM1 patients. It raises the possibility that inter-individual variability in genes related to oxidative stress might contribute to the phenotypic differences observed in these disease. Polymorphism in gene for CYP4F2 and CYP8A1 is related to large lesions count and total lesions, respectively. Intracerebral hemorrhage is associated with CYP46A1 and MMP3 polymorphism. Polymorphism in other oxidative stress genes for glyoxalase and paraoxonase was linked to variation in susceptibility of CCM. These data suggest the potential role of oxidative stress in CCM severity.

Few preclinical reports have been supported beneficial role of antioxidant supplementation in CCM. Tempol is a synthetic molecule to scavenge superoxide in biological system. During CCM drug screening experiments, it inhibited siCCM2 RNA-induced structural–functional phenotypes in endothelial cells, dermal vascular leak, normalized acetylcholine-induced middle cerebral artery vasodilation, and decreased lesion burden in endothelial-specific Ccm2 knockout mice. Tempol did not inhibit upregulated expression of Klf2 and Klf4 in Krit1^{ECKO} cells. Platinum nanozyme (PtNP) is a synthetic monodispersed endotoxin-free nanomaterial with particle sizes 5 and 20 nm. Due to these precise physicochemical characteristics, PtNP is to be more defined for biological properties than other heterogeneous antioxidant nanomaterials such as cerium oxide NP and fullerenes. Functionally, PtNP includes nanocarrier and antioxidant behavior together. Nanocarrier properties lead to accumulation of PtNP inside the endocytic vesicles, and this is without any cytotoxic action on HeLa cells. In terms of antioxidant nature, PtNP possesses SOD, catalase, and peroxidase activities. PtNP decreased ROS stress in the *Krit1* knockout cells. Avenanthramides are secondary metabolites of plants and known for antioxidant and anti-inflammatory activities. Two derivatives of avenanthramide class are generated using bioengineering process from *S. cerevisiae* and tested for the antioxidant activity in Krit1$^{-/-}$MEF cell line. These compounds increased SOD2, FOXO1 expression and reduced cyclin D1 level. Among these agents, tempol is proposed by Ricursion Pharmaceuticals to USFDA for orphan drug designation and achieved approval status in 2015. Currently, Recursion Pharmaceuticals is looking to file IND of tempol for cerebral cavernous malformations.

Except this long pathway of new antioxidants discovery for CCM, FDA-approved antioxidants are available for treatment of neurological diseases, and it is worthier to target them for CCM development. Edaravone (Redicava, 30 mg/100 mL in infusion bag, 2017) is recently approved for long-term treatment of amyotropic lateral sclerosis. Pramipexole (Mirapex ER, 0.375–4.5 mg per tablet) is approved since 1997 for treatment of Parkinson's syndrome. Other than antioxidant, its D2 receptor antagonistic action proposes risk for dyskinesia, hallucinations, and compulsive diseases [71]. Dexpramipexole is devoid of D2 receptor antagonism and tested for amyotropic lateral sclerosis clinical development by Knopp Biosciences and Biogen Idec [71, 72]. In Phase-III EMPOWER trial, dexpramipexole was tolerated but did not improve efficacy with respect to any end points such as combined assessment of function and survival (CAFS) score, amyotrophic lateral sclerosis functional rating scale-revised (ALSFRS-R) total score, and time to death up to 12 months [72]. Currently, it is under clinical development for treatment of hypereosinophilic syndrome by Knopp Biosciences. Edaravone and dexpramipexole are good choice for CCM development.

3.8 Gene Therapy and Local Therapeutic Target Delivery

The pathobiological changes in CCM diseases are related to loss of functional mutation in CCM genes that ultimately lead to dismantling of CCM complex. The restoration of wild-type CCM protein may govern stabilization of CCM complex, reconstruction of endothelial junctions, and barrier integrity. The application of CCM gene therapy will cure the disease. In Phase-I study, the use of adeno-associated virus serotype-2 (AAV2) for direct intracranial delivery of human aspartacylase gene in Canavan disease suffering patients had shown 30% incidences of immunological response by increasing neutralizing antibodies and provided the clinical proof for safety of AAV2 vector in other brain gene delivery [73]. The 2-year follow-up study on stereotactic delivery of AAV2–neuturin (glial cell-derived neurotropic factor) in substantia nigra and putamen regions of Parkinson patients has documented no serious adverse events and no signs of damage on MRI [74]. Another clinical study on Parkinson disease has proven the safety of lentiviral vector and stereotactic surgical procedure for brain gene delivery [75, 76]. However, none of these trials has shown the efficacy by gene delivery method. In murine arteriovenous malformation model, the intravenous administration of AAV2–sFLT02 (a short arm of VEGF receptor) significantly reduced mortality rate and abnormal vessels formation. The Tat peptide-decorated calcitonin gene-related peptide (CGRP) gene containing gelatin–siloxane nanoparticle intracisternal administration had improved neurological functional outcome and decreased vasospasm in murine model of subarachnoid hemorrhage [77]. This non-viral vector delivery system was selective to endothelial cells because of Tat peptide (HIV-1 envelope peptide that requires for cell penetration). These reports are providing insight to conduct preclinical gene therapy trial for CCM management using AAV2 or Tat peptide as a vector- and site-specific intracranial delivery using surgical procedure [77].

Other than these CCM targets-based therapeutic agents, thioram, pyrithione zinc, and silibinin were reported as a selective CCM phenotype (hyperproliferation of astrocytes and EnMT axis, respectively) inhibitors. Gedunin, pindolol, dimercaprol, aloin, and apomorphine have shown anti-CCM actions in drug screening experiments.

4 Conclusion

Taken together, CCM is a rare disease, and the insufficient outcomes by surgical management or evidence-based medications direct the need to develop CCM target-specific pharmacotherapy. The advancement in understanding the molecular mechanisms of CCM pathobiology, preclinical experimental models for drug screening, and clinical outcome monitoring markers for designing trials has opened the avenue to reveal novel pharmacotherapy for CCM. BA-1049, atorvastatin, exisulind, and tempol are currently under CCM preclinical development stage, and none of the drugs are under controlled CCM clinical trial except simvastatin. Marketed FDA-approved drugs (rosuvastatin, paricalcitol, doxercalciferol, ergocalciferol, naloxone, naltrexone, pirfenidone, edaravone, dexpramipexole) and clinically developing drugs for other indications (VS-105, ibudilast, eritoran, TAK-242, galunisertib) are good choice for preclinical development. For clinical development, FDA-approved drugs such as ocrelizumab, ofatumumab and non-FDA-approved drugs such as fasudil are preferential candidates.

Acknowledgements Dr. KKC and Dr. AK summarized this chapter on the basis of the original science reported in the mentioned references. The funders had no role in the study design, data collection, decision to publish, or preparation of the manuscript.

Conflict of Interest The author declares no conflict of interest.

References

1. Akers, A., et al.: Synopsis of guidelines for the clinical management of cerebral cavernous malformations: consensus recommendations based on systematic literature review by the angioma alliance scientific advisory board clinical experts panel. Neurosurgery **80**(5), 665–680 (2017)
2. Gross, B.A., Du, R.: Hemorrhage from cerebral cavernous malformations: a systematic pooled analysis. J. Neurosurg. **126**(4), 1079–1087 (2017)
3. Flemming, K.D., et al.: Population-based prevalence of cerebral cavernous malformations in older adults: Mayo clinic study of aging. JAMA Neurol. **74**(7), 801–805 (2017)
4. Gross, B.A., et al.: The natural history of cerebral cavernous malformations in children. J. Neurosurg. Pediatr. **17**, 1–6 (2015)
5. de Vos, I.J., et al.: Review of familial cerebral cavernous malformations and report of seven additional families. Am. J. Med. Genet. A **173**(2), 338–351 (2017)
6. Gunel, M., et al.: A founder mutation as a cause of cerebral cavernous malformation in Hispanic Americans. N. Engl. J. Med. **334**(15), 946–951 (1996)
7. Laurans, M.S., et al.: Mutational analysis of 206 families with cavernous malformations. J. Neurosurg. **99**(1), 38–43 (2003)
8. Gallione, C.J., et al.: A founder mutation in the Ashkenazi Jewish population affecting messenger RNA splicing of the CCM2 gene causes cerebral cavernous malformations. Genet. Med. **13**(7), 662–666 (2011)
9. Liquori, C.L., et al.: Different spectra of genomic deletions within the CCM genes between Italian and American CCM patient cohorts. Neurogenetics **9**(1), 25–31 (2008)
10. Shenkar, R., et al.: Exceptional aggressiveness of cerebral cavernous malformation disease associated with PDCD10 mutations. Genet. Med. **17**(3), 188–196 (2015)
11. Akers, A.L., et al.: Biallelic somatic and germline mutations in cerebral cavernous malformations (CCMs): evidence for a two-hit mechanism of CCM pathogenesis. Hum. Mol. Genet. **18**(5), 919–930 (2009)
12. Dammann, P., et al.: Correlation of the venous angioarchitecture of multiple cerebral cavernous malformations with familial or sporadic disease: a susceptibility-weighted imaging study with 7-Tesla MRI. J. Neurosurg. **126**(2), 570–577 (2017)
13. Oka, M., et al.: Therapeutic potential of RhoA/Rho kinase inhibitors in pulmonary hypertension. Br. J. Pharmacol. **155**(4), 444–454 (2008)
14. Yao, L., et al.: The role of RhoA/Rho kinase pathway in endothelial dysfunction. J. Cardiovasc. Dis. Res. **1**(4), 165–170 (2010)
15. Bond, L.M., Sellers, J.R., McKerracher, L.: Rho kinase as a target for cerebral vascular disorders. Future Med. Chem. **7**(8), 1039–1053 (2015)

16. Shibuya, M., et al.: Effects of fasudil in acute ischemic stroke: results of a prospective placebo-controlled double-blind trial. J. Neurol. Sci. **238**(1–2), 31–39 (2005)
17. Shimokawa, H., et al.: Anti-anginal effect of fasudil, a Rho-kinase inhibitor, in patients with stable effort angina: a multicenter study. J. Cardiovasc. Pharmacol. **40**(5), 751–761 (2002)
18. Shibuya, M., et al.: Effect of AT877 on cerebral vasospasm after aneurysmal subarachnoid hemorrhage. Results of a prospective placebo-controlled double-blind trial. J. Neurosurg. **76**(4), 571–577 (1992)
19. Fukumoto, Y., et al.: Double-blind, placebo-controlled clinical trial with a rho-kinase inhibitor in pulmonary arterial hypertension. Circ. J. **77**(10), 2619–2625 (2013)
20. Jiang, X., et al.: Acute hemodynamic response of infused fasudil in patients with pulmonary arterial hypertension: a randomized, controlled, crossover study. Int. J. Cardiol. **177**(1), 61–65 (2014)
21. Suzuki, Y., et al.: A postmarketing surveillance study of fasudil treatment after aneurysmal subarachnoid hemorrhage. Surg. Neurol. **68**(2), 126–131 (2007). (discussion 131–2)
22. Guo, R., et al.: Fasudil improves short-term echocardiographic parameters of diastolic function in patients with type 2 diabetes with preserved left ventricular ejection fraction: a pilot study. Heart Vessels **30**(1), 89–97 (2015)
23. Otsuka, T., et al.: Administration of the Rho-kinase inhibitor, fasudil, following nitroglycerin additionally dilates the site of coronary spasm in patients with vasospastic angina. Coron. Artery Dis. **19**(2), 105–110 (2008)
24. McDonald, D.A., et al.: Fasudil decreases lesion burden in a murine model of cerebral cavernous malformation disease. Stroke **43**(2), 571–574 (2012)
25. Shenkar, R., et al.: RhoA kinase inhibition with fasudil versus simvastatin in murine models of cerebral cavernous malformations. Stroke **48**(1), 187–194 (2017)
26. Bedi, O., et al.: Pleiotropic effects of statins: new therapeutic targets in drug design. Naunyn Schmiedebergs Arch. Pharmacol. **389**(7), 695–712 (2016)
27. Schachter, M.: Chemical, pharmacokinetic and pharmacodynamic properties of statins: an update. Fundam. Clin. Pharmacol. **19**(1), 117–125 (2005)
28. Gresser, U., Gathof, B.S.: Atorvastatin: gold standard for prophylaxis of myocardial ischemia and stroke—comparison of the clinical benefit of statins on the basis of randomized controlled endpoint studies. Eur. J. Med. Res. **9**(1), 1–17 (2004)
29. Silva, M., et al.: Meta-analysis of drug-induced adverse events associated with intensive-dose statin therapy. Clin. Ther. **29**(2), 253–260 (2007)
30. Quinn, K.L., et al.: Lipophilic statins and the risk of intracranial hemorrhage following ischemic stroke: a population-based study. Drug Saf. (2017)
31. Glueck, C.J., et al.: Rosuvastatin 5 and 10 mg/d: a pilot study of the effects in hypercholesterolemic adults unable to tolerate other statins and reach LDL cholesterol goals with nonstatin lipid-lowering therapies. Clin. Ther. **28**(6), 933–942 (2006)
32. Manoukian, G.E., et al.: Pilot trial of bone-targeted therapy combining zoledronate with fluvastatin or atorvastatin for patients with metastatic renal cell carcinoma. Clin. Genitourin. Cancer **9**(2), 81–88 (2011)
33. Shah, S., Savjani, J.: A review on ROCK-II inhibitors: from molecular modelling to synthesis. Bioorg. Med. Chem. Lett. **26**(10), 2383–2391 (2016)
34. Bittner, S., et al.: Targeting B cells in relapsing-remitting multiple sclerosis: from pathophysiology to optimal clinical management. Ther. Adv. Neurol. Disord. **10**(1), 51–66 (2017)
35. Lim, S.H., et al.: Anti-CD20 monoclonal antibodies: historical and future perspectives. Haematologica **95**(1), 135–143 (2010)
36. Kappos, L., et al.: Ocrelizumab in relapsing-remitting multiple sclerosis: a phase 2, randomised, placebo-controlled, multicentre trial. Lancet **378**(9805), 1779–1787 (2011)
37. Sorensen, P.S., et al.: Safety and efficacy of ofatumumab in relapsing-remitting multiple sclerosis: a phase 2 study. Neurology **82**(7), 573–581 (2014)
38. Hauser, S.L., et al.: Ocrelizumab versus interferon beta-1a in relapsing multiple sclerosis. N. Engl. J. Med. **376**(3), 221–234 (2017)

39. Montalban, X., et al.: Ocrelizumab versus placebo in primary progressive multiple sclerosis. N. Engl. J. Med. **376**(3), 209–220 (2017)
40. Reagan, P.M., Friedberg, J.W.: Reassessment of anti-CD20 therapy in lymphoid malignancies: impact, limitations, and new directions. Oncology (Williston Park) **31**(5), 402–411 (2017)
41. Frampton, J.E.: Ocrelizumab: first global approval. Drugs **77**(9), 1035–1041 (2017)
42. Troletti, C.D., et al.: Molecular alterations of the blood-brain barrier under inflammatory conditions: the role of endothelial to mesenchymal transition. Biochim. Biophys. Acta **1862**(3), 452–460 (2016)
43. Maddaluno, L., et al.: EndMT contributes to the onset and progression of cerebral cavernous malformations. Nature **498**(7455), 492–496 (2013)
44. Sinibaldi, V.J., et al.: Phase II evaluation of docetaxel plus exisulind in patients with androgen independent prostate carcinoma. Am. J. Clin. Oncol. **29**(4), 395–398 (2006)
45. Weight, C.J., et al.: A prospective, controlled phase II study of neoadjuvant exisulind therapy before radical prostatectomy: effect on apoptosis. Urology **80**(2), 484-e17–22 (2012)
46. Attia, S., et al.: Phase I/II study of vinorelbine and exisulind as first-line treatment of advanced non-small cell lung cancer in patients at least 70 years old: a wisconsin oncology network study. J. Thorac. Oncol. **3**(9), 1018–1025 (2008)
47. Weiss, G.J., et al.: Docetaxel and exisulind in previously treated non-small cell lung cancer (NSCLC) patients: a multicenter, phase II clinical trial. J. Thorac. Oncol. **2**(10), 933–938 (2007)
48. Pusztai, L., et al.: Phase I and II study of exisulind in combination with capecitabine in patients with metastatic breast cancer. J. Clin. Oncol. **21**(18), 3454–3461 (2003)
49. King Jr., T.E., et al.: A phase 3 trial of pirfenidone in patients with idiopathic pulmonary fibrosis. N. Engl. J. Med. **370**(22), 2083–2092 (2014)
50. Herbertz, S., et al.: Clinical development of galunisertib (LY2157299 monohydrate), a small molecule inhibitor of transforming growth factor-beta signaling pathway. Drug Des. Devel. Ther. **9**, 4479–4499 (2015)
51. Nathan, S.D., et al.: Effect of continued treatment with pirfenidone following clinically meaningful declines in forced vital capacity: analysis of data from three phase 3 trials in patients with idiopathic pulmonary fibrosis. Thorax **71**(5), 429–435 (2016)
52. Brandes, A.A., et al.: A Phase II randomized study of galunisertib monotherapy or galunisertib plus lomustine compared with lomustine monotherapy in patients with recurrent glioblastoma. Neuro Oncol. **18**(8), 1146–1156 (2016)
53. Capper, D., et al.: Biomarker and histopathology evaluation of patients with recurrent glioblastoma treated with galunisertib, lomustine, or the combination of galunisertib and lomustine. Int. J. Mol. Sci. **18**(5), 995 (2017)
54. Ikeda, M., et al.: Phase 1b study of galunisertib in combination with gemcitabine in Japanese patients with metastatic or locally advanced pancreatic cancer. Cancer Chemother. Pharmacol. **79**(6), 1169–1177 (2017)
55. Cao, Y., et al.: NGS-based transcriptome profiling reveals biomarkers for companion diagnostics of the TGF-beta receptor blocker galunisertib in HCC. Cell Death Dis. **8**(2), e2634 (2017)
56. Chen, Q., et al.: Elevated cerebrospinal fluid levels of thrombospondin-1 correlate with adverse clinical outcome in patients with aneurysmal subarachnoid hemorrha*ge*. J. Neurol. Sci. **369**, 126–130 (2016)
57. Dong, X.Q., et al.: Changes in plasma thrombospondin-1 concentrations following acute intracerebral hemorrhage. Clin. Chim. Acta **450**, 349–355 (2015)
58. Shen, Y.F., et al.: The prognostic value of plasma thrombospondin-1 concentrations after aneurysmal subarachnoid hemorrhage. Clin. Chim. Acta **448**, 155–160 (2015)
59. Baker, L.H., et al.: Randomized, phase II study of the thrombospondin-1-mimetic angiogenesis inhibitor ABT-510 in patients with advanced soft tissue sarcoma. J. Clin. Oncol. **26**(34), 5583–5588 (2008)
60. Sahora, A.I., et al.: Prospective study of thrombospondin-1 mimetic peptides, ABT-510 and ABT-898, in dogs with soft tissue sarcoma. J. Vet. Intern. Med. **26**(5), 1169–1176 (2012)

61. Hoekstra, R., et al.: Phase I safety, pharmacokinetic, and pharmacodynamic study of the thrombospondin-1-mimetic angiogenesis inhibitor ABT-510 in patients with advanced cancer. J. Clin. Oncol. **23**(22), 5188–5197 (2005)
62. Gietema, J.A., et al.: A phase I study assessing the safety and pharmacokinetics of the thrombospondin-1-mimetic angiogenesis inhibitor ABT-510 with gemcitabine and cisplatin in patients with solid tumors. Ann. Oncol. **17**(8), 1320–1327 (2006)
63. Hoekstra, R., et al.: Phase I study of the thrombospondin-1-mimetic angiogenesis inhibitor ABT-510 with 5-fluorouracil and leucovorin: a safe combination. Eur. J. Cancer **42**(4), 467–472 (2006)
64. Gordon, M.S., et al.: A phase 1 trial of 2 dose schedules of ABT-510, an antiangiogenic, thrombospondin-1-mimetic peptide, in patients with advanced cancer. Cancer **113**(12), 3420–3429 (2008)
65. Nabors, L.B., et al.: A phase 1 trial of ABT-510 concurrent with standard chemoradiation for patients with newly diagnosed glioblastoma. Arch. Neurol. **67**(3), 313–319 (2010)
66. Molckovsky, A., Siu, L.L.: First-in-class, first-in-human phase I results of targeted agents: highlights of the 2008 American society of clinical oncology meeting. J. Hematol. Oncol. **1**, 20 (2008)
67. Li, L., et al.: Antitumor efficacy of a thrombospondin 1 mimetic CovX-body. Transl. Oncol. **4**(4), 249–257 (2011)
68. Najjar, S., et al.: Neurovascular unit dysfunction and blood-brain barrier hyperpermeability contribute to schizophrenia neurobiology: a theoretical integration of clinical and experimental evidence. Front Psychiatry **8**, 83 (2017)
69. Imai, T., et al.: Nrf2 activator ameliorates hemorrhagic transformation in focal cerebral ischemia under warfarin anticoagulation. Neurobiol. Dis. **89**, 136–146 (2016)
70. Laplace, C., et al.: Endothelial oxidative stress induced by serum from patients with severe trauma hemorrhage. Intensive Care Med. **31**(9), 1174–1180 (2005)
71. Pandya, R.S., et al.: Therapeutic neuroprotective agents for amyotrophic lateral sclerosis. Cell. Mol. Life Sci. **70**(24), 4729–4745 (2013)
72. Cudkowicz, M.E., et al.: Dexpramipexole versus placebo for patients with amyotrophic lateral sclerosis (EMPOWER): a randomised, double-blind, phase 3 trial. Lancet Neurol. **12**(11), 1059–1067 (2013)
73. McPhee, S.W., et al.: Immune responses to AAV in a phase I study for Canavan disease. J. Gene. Med. **8**(5), 577–588 (2006)
74. Warren Olanow, C., et al.: Gene delivery of neurturin to putamen and substantia nigra in Parkinson disease: a double-blind, randomized, controlled trial. Ann. Neurol. **78**(2), 248–257 (2015)
75. Bartus, R.T., et al.: Safety/feasibility of targeting the substantia nigra with AAV2-neurturin in Parkinson patients. Neurology **80**(18), 1698–1701 (2013)
76. Palfi, S., et al.: Long-term safety and tolerability of ProSavin, a lentiviral vector-based gene therapy for Parkinson's disease: a dose escalation, open-label, phase 1/2 trial. Lancet **383**(9923), 1138–1146 (2014)
77. Tian, X.H., et al.: Tat peptide-decorated gelatin-siloxane nanoparticles for delivery of CGRP transgene in treatment of cerebral vasospasm. Int. J. Nanomed. **8**, 865–876 (2013)

MIX
Papier aus verantwortungsvollen Quellen
Paper from responsible sources
FSC® C105338

If you have any concerns about our products,
you can contact us on
ProductSafety@springernature.com

In case Publisher is established outside the EU,
the EU authorized representative is:
Springer Nature Customer Service Center GmbH
Europaplatz 3, 69115 Heidelberg, Germany

Printed by Libri Plureos GmbH
in Hamburg, Germany